The Patrick Moore Practical Astronomy Series

More information about this series at http://www.springer.com/series/3192

Deep Sky Observing

An Astronomical Tour

Steven R. Coe

Second Edition

Steven R. Coe
Livingston, TX, USA

ISSN 1431-9756 ISSN 2197-6562 (electronic)
The Patrick Moore Practical Astronomy Series
ISBN 978-3-319-22529-6 ISBN 978-3-319-22530-2 (eBook)
DOI 10.1007/978-3-319-22530-2

Library of Congress Control Number: 2015951792

Springer Cham Heidelberg New York Dordrecht London

Cover photo credit: Courtesy of Grand Canyon National Park Service. Used with a Creative Commons license. Image taken from the Grand Canyon NPS annual star party

Printed on acid-free paper

Springer International Publishing AG Switzerland is part of Springer Science+Business Media (www.springer.com)

Acknowledgments

This book is dedicated to the members of the Saguaro Astronomy Club in Phoenix, Arizona, USA (See Fig. 1).

I acquired the ability to write about deep-sky observing by "practicing" in the club newsletter. I treasure the years of their fellowship while observing the Universe.

Thanks to all.

Steven R. Coe
2016

Fig. 1 The Saguaro Astronomy Club set up and ready for the Messier Marathon

Fig. 2 Saguaro Astronomy Club at a star party in the cool pines

Contents

Chapter 1

Who Can Benefit from This Book?

Before starting this book I realized that there are lots of astronomy books that are written to help beginners get started with observing the sky. It is important for you to know that this is *not* the traditional text that is available to an astronomy novice. If you know absolutely nothing about viewing the sky, then this is not a book for you.

I have listed some excellent books that are useful for novice astronomers in Chap. 18; please choose one of those and read it first. This book will not include the definition of a "light year", nor will it provide information about the difference between reflectors and refractors. There is very little information on which telescope to buy; these subjects are explained quite well in the recommended beginner's books. What *will* be explained is how to use that equipment to view and *enjoy* a variety of deep-sky objects.

So, you own a star chart and can find your way around it well enough to see the brightest objects. Your next question is likely to be, "What else is out there that I can look at?" The short answer is: a lot. There is much more to see of the Universe.

If you are already at that level of sophistication and want to find out more about observing deep-sky objects with your telescope, then this is the book for you. I will cover a wide variety of questions that members of my astronomy club have asked over the years.

I have been the chairman of the Novice Group within the Saguaro Astronomy Club of Phoenix, Arizona, for over 20 years and have been called upon to answer many questions about astronomy—questions like these:

"What books do I need?"
"How can my computer search for information about deep-sky objects?"
"Should I get the big star charts?"
"What eyepieces should I buy?"

S.R. Coe, *Deep Sky Observing*, The Patrick Moore Practical Astronomy Series,
DOI 10.1007/978-3-319-22530-2_1

Fig. 1.1 Steve Coe and 13 in.

"Is this how NGC 7789 should look with an 8 inch telescope?"
"What is averted vision, and when do I use it?"
"How far do I need to drive away from the city to get dark skies?"
"What will a larger telescope do for my viewing?"
"What will a larger telescope do for my lower back pain?"
"What will a larger telescope do for my savings account?"

Some of these are pretty difficult questions, but I know that I can be helpful in answering them.

Most importantly, I want this book to be about the *fun* of observing the Universe in all its splendor. Taking the time at the end of a hectic day to set up the telescope and view a star cluster that very few other people have ever seen will always give you a sense of enjoyment—not that this is a race or a tournament. It's a chance to view the largest panorama of all.

And remember that astronomy is really supposed to be a joyous hobby, not a second job!

I saw a message on the Internet from an observer whose name was given as just "Mitch". The message said that there are two kinds of amateur astronomers: (1) Those who think that a telescope is a means to gaze at the heavens and (2) Those who think that the heavens create a need to build a telescope. Even though I am a member of the second group, I think we are fortunate to have both approaches: it makes for a fun mix of people at star parties.

Fig. 1.2 On our way to the Grand Canyon. Left to right: Bill Anderson, Dave Frederickson, A.J. Crayon and Steve Coe

As the chapters of this book progress, I will be mentioning several of my friends. These folks and I have been observing together for over 20 years and they figure prominently in some of the tales that are told later in this book (See Fig. 1.2).

Bill Anderson has a 6 in. (150 mm) f/5 as a rich field telescope for wide field viewing and a Celestron 8 in. (200 mm) SCT. Bill has excellent computer skills and a great deal of expertise in telescope construction.

A.J. Crayon has an 8 in. (200 mm) SCT. A.J. has been a member of the Saguaro Astronomy Club for 20 years and is the Chairman of the Deep Sky Group. He also is the coordinator of the Arizona Messier Marathon, one of the most prolific Messier searches in the world (See Fig. 1.3).

David Fredericksen passed on several years ago and I miss him. David was a SAC member for 20 years. David taught physics and science at a local high school and has taught astronomy classes at Glendale College (See Fig. 1.4).

Chris Schur is an excellent astrophotographer. His work has appeared in a wide variety of astronomy books and magazines. Some of the photos in this book are by Chris (See Fig. 1.5).

Fig. 1.3 A.J. observing Comet Hale–Bopp with his 8 in. f/6 Newtonian

Fig. 1.4 Dave and his 12½ in. f/6 Newtonian

Fig. 1.5 Chris Schur

Pierre Schwaar has also passed away and I miss him as well. He was a master telescope maker and mirror grinder. His mirrors are legendary in Arizona and elsewhere. My 13 in. (330 mm) f/5.6 is a creation from Pierre's workshop (See Fig. 1.6).

I have learned much from these people and others. It has been a fascinating and enlightening journey.

So, 15 years have passed since the first edition of this book was published. I have had some good feedback and have changed some things in this version. I will try and get the dark images to brighten up so that a reader can figure out what was supposed to be in the picture. Several folks told me that they would like to read some observations with smaller telescopes. I have added text that addresses that request. I owned a Televue TV 102 refractor for several years and I observed a lot with that 4 in. scope. It was a fine instrument, but I found that in a couple of years I managed to view many deep sky objects and anything fainter was not worth observing with that small an aperture, in my opinion. But bright and large objects really were beautiful in that excellent refractor.

I have had a chance to view and image the night sky over the past 15 years. And I have learned much about being an amateur astronomer. For the past 6 years I have been living in my motor home and enjoyed very much the chance to travel the

Fig. 1.6 Pierre Schwaar drawing sunspots

western United States. I have particularly enjoyed the Oregon Star Party as my favorite of the big gatherings of amateurs. There is plenty to see in the sky and fun people to chat with while doing that.

Closer to home in Arizona we have two star parties each year. Both the All Arizona Star Party and the All Arizona Messier Marathon are well attended and provide a chance to meet with old friends and make new ones. The Messier Marathon is in the Spring and the big All Arizona Star Party in the Autumn so there are different parts of the sky to enjoy. Arizona is a state that is filled up with telescopes, both permanent observatories and mobile. It is a great place to enjoy the night sky.

Having the motor home available is great; there is a real bed, bathroom, microwave oven and furnace to make staying in the dark Arizona desert a more comfortable experience. When the summer heat comes along I can drive north to the cool pines and view from there. There are several members of the Saguaro Astronomy Club who have told me that they are jealous.

As I write this, I am two weeks away from my 66th birthday—how did that happen? It has been a grand life, I have traveled, loved a woman, stayed 26 years at a job that I enjoyed very much and have been pretty healthy while doing all of that. I certainly see how lucky I am to have all this and astronomy too.

I have had email, private messages from the Cloudy Nights website and a few phone calls from people getting in touch to say how much they enjoyed the first edition of this book. I am so happy to hear from satisfied "customers". And I have had some feedback on how to make it better. I hope to address that in this version. My email address is stevecoe at cloudynights dot com. Let me know what you think.

Next year (2016) I will be an Arizona amateur astronomer for 40 years. It has been a great time. I have friends that I have been observing with for most of that time. It has added so much to the enjoyment of viewing the sky to share it with others. There have been times when I got hooked on imaging the wide field sky; this is one of those times. Most of the time I spent visually observing and sketching what I saw at the eyepiece. I have no doubt I will return to observing full time when I have finished imaging the objects on my list.

Regardless of having viewed the sky for virtually all of my adult life, I have never grown tired of it. A trip to dark skies allows me to set up the scope, sit in a chair and enjoy the view as the Sun sets and stars come out under a clear sky. We do live in a beautiful universe.

So now that you've read this far, let's proceed. Each of the chapters is the answer to a question that some-one has asked before.

Chapter 2

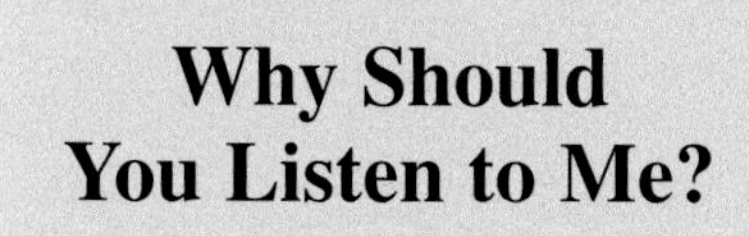

Why Should You Listen to Me?

I have been a deep-sky observer for over 30 years, almost all of that while living in Arizona. I am perfectly clear how fortunate that makes me. The state of Arizona is sprinkled with astronomical observatories of all types. Many observers of the sky move to this part of the country so they can take advantage of the clear desert skies. While in the U.S. Navy I had the opportunity to live in the states of Washington and New York. These are places where the weather and city lights make astronomy difficult and often frustrating (See Fig. 2.1).

Being an amateur astronomer for many years provides an opportunity to make lots of mistakes and I have made plenty. Also, having been the Novice Group coordinator for many years gives me an appreciation for the frustration facing a beginner who is finding their way among the confusing sky. Many of the members of my club have said that I appreciate their problems and can often provide some input that is helpful, and this is one of the reasons I am writing this book.

There is much to know and understand on your path toward having fun while observing the deep sky. Before you can see the joy of observing a beautiful edge-wise galaxy, you have to find it and put it in the field of view of your telescope! I realize how obvious this statement seems, but finding your way around the sky is far from easy. My purpose is to provide helpful information, and assist you with some of the expertise I have discovered while treading the same path. That way, you can get to the fun of seeing the beauty of the night sky more quickly.

A wide variety of telescopes and other astronomy devices have been part of my observing life, and yet the first nugget of wisdom is this: don't spend lots of money, time and mental effort concerning yourself about astronomy equipment. So often a newcomer to observing will spent lots of money for equipment that they realize later is not necessary. Take it easy; the sky will still be there. You don't have to own

S.R. Coe, *Deep Sky Observing*, The Patrick Moore Practical Astronomy Series,
DOI 10.1007/978-3-319-22530-2_2

Fig. 2.1 32 in. telescope and milky way

a huge collection of eyepieces, star charts or filters to enjoy the Universe. There is certainly a chapter in this book that launches into a discussion of accessories. But I hope you'll listen to me and won't fall into the very enticing trap of "equipment collecting".

My journey as an amateur astronomer has been anything but smooth. I know that much of what I've learned is generated by not making the same mistake twice. The fairly obvious point I made in the previous paragraph—not to be overly concerned about equipment—was learned from years of purchasing the latest gadget only to find it is *not* the Swiss Army knife of astronomy. The good news is that you can usually find someone gullible enough to purchase this device for about 80 % of its original price… The bad news is that you will quickly spend that money on yet another gadget if you don't take the time to consider "What do I really need?"

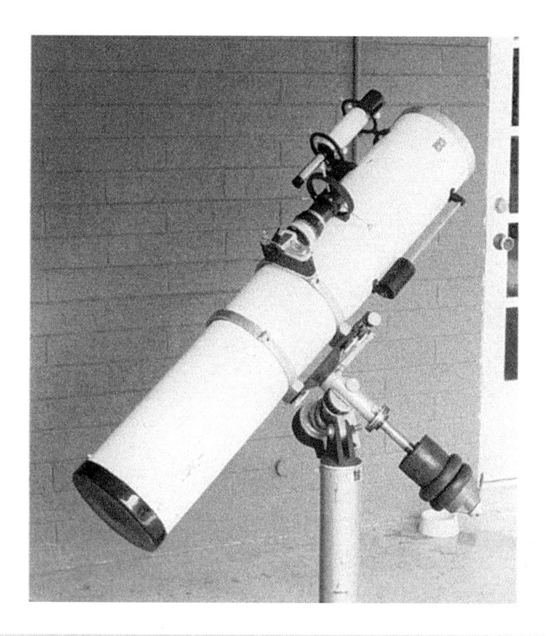

Fig. 2.2 8 in. f/6—my first scope

Over the past 5 years or so, I have rid myself of some rarely used accessories, which has greatly simplified my observing. I have also discovered more room in my garage and my storage closet.

Knowing more about me might help you see that I've waded through many of the same astronomical swamps that you are facing. I started out at Arizona State University to be a professional astronomer. However, I found out quickly that I did not want to be a mathematician for the rest of my life. I want to *observe* the sky, not measure it! I did manage to get a degree in communications and found the job I had from 1980 to 2006. I taught electronics at DeVry Institute of Technology in Phoenix, Arizona. Fortunately, it provided enough money (and time off) for astronomical observing.

My first scope was an 8 in. (200 mm) f/6 Newtonian and I thought I wanted to be an astrophotographer. My first attempt had me trying to sit down to guide the picture, then my knee hit the declination shaft and knocked the entire scope over onto its side. It came to me that I just wanted to look; and so I did. After 2 years of observing with the 8 in. telescope, the aperture bug bit me and a 17.5 in. (450 mm) mirror was ordered. Several scopes were constructed around that mirror and many observations were made with those telescopes (See Fig. 2.3).

However, I missed having a tracking mount. So I sold the big mirror and went to a smaller aperture with a German equatorial mount that is motorized. In the Saguaro Astronomy Club we were fortunate to have Pierre Schwaar, a telescope maker of extraordinary skill. He ground and polished a 13 in. (330 mm) f/5.6 mirror

Fig. 2.3 17½ in. and a younger me

Fig. 2.4 Tom and Jeannie Clark's 36 in. f/5 will give anyone aperture envy

and constructed a mount to fit it. I added a few accessories, which I will discuss in detail later. Most of the observations in this book were made with this telescope. I also have a 6 in. (150 mm) f/6 Newtonian, which is a wonderful rich-field telescope (RFT). It provides beautiful wide-field views of the sky.

During the time between the first and second editions of this book, I have owned several other telescopes. I had a refractor binge and owned a 3 in. (80 mm) ED 80, a 4 in. (102 mm) TV 102 and a 6 in. (150 mm) Celestron C6R. These telescopes have allowed me to provide some observations with smaller aperture.

My good fortune in meeting some well-known observers with large telescopes has provided me some excellent opportunities to observe with large apertures. I wrote articles for *Amateur Astronomy* magazine for several years. The editors, Tom and Jeannie Clark, have a 36 in. (910 mm) f/5 telescope that has provided me with excellent views using a large light collector. On a trip to Texas, Larry Mitchell was kind enough to allow Brian Skiff and me to use his 36 in. f/5 for several nights. My sincere thanks to all the friends who have been kind enough to allow me a peek through their telescopes (See Fig. 2.4).

I do some simple astrophotography using a digital single lens reflex cameras that provide wide field views of the sky. Some of my photos are in this book along with those of other dedicated astrophotographers. All the drawings are mine.

Chapter 3

How Do I Find the Best Observing Site?

The first step in observing the deep sky is finding an observing site.

If you are like most astronomers, you are going to have to put up with some city lights at your "dark-sky" site. *Drive to an area which will put the darkest sky in the direction you plan to observe.* The best way to begin looking for a site is get a good map of the area and determine where the lights are the most problem. There are several poster-size photos of the "continent at night". I have seen them for both North America and Europe. Spend some time with those satellite photos and a map, and then you can spot all the large, bright cities in an effort to avoid them.

One thing that seems obvious is that you must make an observing list for the part of the sky which will be dark at your particular observing site.

For instance, if you are driving to the east, do not make up an observing list of objects that are setting in the west. You will have driven for an hour and set up the scope, only to be chasing objects that are lost in the city lights.

In general, make an observing list that is in the sky direction you are driving toward. So if you are driving to the east, make a list of galaxies, nebulae and clusters that are well above the eastern horizon. That way, they will be "in the dark" once the Sun goes down. All of this is assuming that you only have one light dome that you are concerned about. If there are multiple cities nearby, you just have to do the best you can. Look at your maps carefully and minimize the sky brightness of your site.

Just looking at your new map, any reasonable person would think that the system of national parks in America is ready-made for astronomy. So did we. A 2 h drive from Phoenix led us to think that we would be all alone in a small park. When we arrived, there wasn't enough room left in that park to set up a 60 mm refractor! Any park which is on a road map is generally going to be used heavily when the weather

S.R. Coe, *Deep Sky Observing*, The Patrick Moore Practical Astronomy Series,
DOI 10.1007/978-3-319-22530-2_3

Fig. 3.1 Getting far away from the city lights allows the Milky Way to really show off

is good and crowds are looking to get out of the city. Like it or not, those crowds are just like you (See Fig. 3.1).

What seems to be the best site is just a dark, quiet country road. You may have to determine if the land is privately owned, but we have had pretty good luck. A conversation with the owners so that they see that really all your group is going to do is set up telescopes and observe the sky will often get you an invitation to use the property.

Occasionally, we drive to a new site to find that the lights of some unexpected source are interfering with what we wanted to observe.

This is particularly a problem with comets or other objects which are often near the horizon where the lights can be most bothersome. If you don't need to get too close to the horizon, then try using hills or trees to block the unwanted light source. A vehicle can also be used to block off lights if it is flat country.

There are some things which cannot be changed. In the local hunting season lots of people are going to be out in the woods. Be careful! This may be a good time to stay near a well-known park and put up with some lights from fires and camp

Fig. 3.2 A Celestron Nexstar 11 pointed at a field of stars, the observing site is 60 miles from Phoenix

lanterns rather than be mistaken for prong-horned antelope. Another problem is that outdoor parties are the favorite fun of lots of teenagers. On several occasions car-loads of young people have driven by with headlights ablaze. A well-lit telescope is a sad thing to see. Just find a new location when this sort of thing happens too often.

And don't drive to a new site on a cherished New Moon Saturday night and expect to stumble on the best location. Either show up well before sunset and take some time to scout around or drive to the area on a Third-Quarter Moon evening. That way, you can avoid setting up near a well-traveled road or overlooking a brightly lit football field.

You are probably going to be driving late at night on back roads, so several precautions seem to be in order. Travelling alone may mean that you will have to walk for help. Having several vehicles at a remote site can reduce the effect of a wide variety of problems.

Fig. 3.3 Trucking the telescope far from town can provide dark skies and memorable nights

Make certain that the spare tire is in good condition and inflated or carry an inflation device. Carry several quarts of oil and a gallon or two of water or anti-freeze for the radiator. A few flares, a set of jumper cables, a blanket and first-aid kit will help in case of an emergency. If you are very cautious, a spare fan belt and some fuses to fit the car could make a big difference. Duct tape (gaffer tape) can be used for a variety of purposes, including repair of a leaking hose (See Fig. 3.3).

Often we have several cars and therefore like to chat while we are driving to the site. A CB radio will make the trip go faster and safer. Whoever is in the lead auto can inform the others of upcoming problems. It also helps to keep everyone awake during the trip home when it is late at night. Choose a channel away from the large volume of radio traffic on Channels 19 or 21 (and 14 in the UK). Prices have come down over the last few years and a good CB rig can be had for $100 or less.

And don't forget that, especially in Europe (and that includes the UK) even "remote" sites may be within cell-phone range. Take one along.

Lots of people have a wonderful memory of camping in the woods and how the stars lit up the mountain sky. However, it seems that in practice going too high can be a problem. It might appear that if getting to 5000 ft (1500 m) of altitude is good, then climbing above 8000 ft (2500 m) ought to be great. My trips to extremely high altitudes have not been good ones, however. It seems that the lack of oxygen at altitudes above 7000 ft (2200 m) starts to degrade human vision. I have never seen stars to a fainter magnitude from a very high altitude (See Fig. 3.4).

It is definitely worth the effort to travel to dark skies. There are plenty of obstacles to driving away from the city for a weekend of observing. But the views of the Universe that this precious time provides will stay with you for a lifetime.

From within the city of Phoenix, only the bright stars of the "Teapot" in Sagittarius can be picked out, and then only if the sky is transparent. At a site that I generally rate 5 out of 10 for contrast and transparency, the "Teapot" asterism is

Fig. 3.4 Mountain skies are worth the trip

obvious and the brighter parts of the Milky Way are pretty easily visible. These locations are an hour's drive from home. However, driving 2 h away from Phoenix to a truly dark and transparent site will transform the constellation of Sagittarius into an amazing light show. The Milky Way glows like a beacon and bright spots within Our Galaxy show off dozens of star clusters and gas clouds, all shouting "Look at ME!" (See Fig. 3.4)

Chapter 4

How Do I Maximize My Time While Observing?

Most people observe from their backyards and save up trips to dark-sky sites for New Moon weekends. My friends and I do the same thing. This chapter is going to provide some information that will help you avoid wasting any of those cherished hours under dark skies.

Be Weather-Wise

Becoming a deep-sky observer will also make you become a weather watcher. Get a group of amateur astronomers together and eventually, just like with farmers, the subject of conversation will become "the weather". Will it be clear tonight? Is that some high cirrus cloud moving in? Did you look at the weather report before we left? (See Fig. 4.1)

Pay attention to clouds and cloud types; they have an important story to tell. The weather man calls the large, cotton ball clouds *fair weather cumulus* because they are generally not the precursors of storms. Thick, layered clouds are called *stratus* because of the layers. These clouds will shut down astronomy for days or weeks as they continue to appear and will cover the entire sky. This is a good time to make an observing list and clean your optics! *Cirrus* clouds are the high, thin clouds that can dissipate after sunset, when the Sun's energy is no longer generating them (See Fig. 4.2).

Your local TV or radio news broadcast or newspaper will generally have a weather map, along with a satellite picture of your portion of the country. Many Internet sites offer a weather service, with regional maps and forecasts. In the UK the (free) *teletext* services also provide up-to-date regional weather forecasts.

S.R. Coe, *Deep Sky Observing*, The Patrick Moore Practical Astronomy Series,
DOI 10.1007/978-3-319-22530-2_4

Fig. 4.1 A telescope set up under cloudy skies is a sad thing to see

Fig. 4.2 Here we are, all set up at the Grand Canyon and ready for a great night. We checked the weather carefully before this long drive

Satellite images can show cloud movement better than a surface map. Watch the movement of storms and pressure ridges; they will start to show a pattern over an extended period. It takes several days after a large storm for the chaotic movement of the atmosphere to settle down and provide good seeing. It will take several years to get good at predicting cloud cover, but you should get started. To put it simply, pay attention and you will see a set of conditions that will repeat itself over time. At most locations, you will see several weather conditions that show up again and again.

As an example, in the Southwestern United States there is always some dust in the air. You can learn to predict how much it will affect the viewing by looking at the color of the daytime sky. A white or light-blue sky means that there is lots of dust scattering the sunlight in the atmosphere. This is a good night to forget the faint objects on your observing list and go after the brighter deep sky objects. However, when the sky is medium to dark blue then the dust is minimal and you can expect to chase those faint, elusive nebulae.

Learn to Use Your Vision to the Best of Your Ability

When we get to the chapter on taking notes there will be lots of information about how to write down what you see. Before that happens, you must learn to use your eyes and mind to the best of their ability.

The daytime sky is over 50 million times brighter than the night-time sky and yet your eyes work under both conditions. Two changes to your vision make this possible. The first is that your eyes' pupils will open wider as the night approaches. Also, a chemical called visual purple changes your retina so that it is ready to accept dimmer lighting conditions.

These changes take an hour or two, but can be quickly reversed by a poorly controlled white light flashlight. Once you get dark-adapted you must make certain that you stay in that condition by not allowing bright lights to cancel your dark adaptation. Here is a simple experiment which will prove to you how dark-adapted you are after 3 or 4 h under dark skies.

Make certain either that you are observing alone or that everyone is finished observing. Now, turn on a white flashlight and read a page of a book. This will completely cancel your dark-adapted vision. Then turn off the flashlight and see how dark your observing site appears. The difference is all in your eye.

A technique for seeing all the detail available within a galaxy or nebula is to use *averted vision*. In the past you might have noticed something "out of the corner of your eye". Well, that (the corner of your eye) is what's meant by "averted vision".

The center of your field of view is the place on your retina with the sharpest and most detailed vision, but it *isn't* the part that is most sensitive to light. Look to the side of the field a little and you might well see some faint companion or detail that with direct vision you would miss.

Keep Warm

First, be comfortable. Take plenty of warm clothes. A full ski suit will provide complete body coverage and doesn't have gaps which let your back or neck get cold. Use a muffler, mask and ski cap to keep your head warm. Warm socks and moon boots help a lot to keep your feet warm. Many different types of insulated boots are available; try them on for a while and remember you will be standing in

Fig. 4.3 Warm boots are a must on a cool night

them for many an hour when at the telescope. Take a jacket, a pair of thick pants and a cap even when it is a warm day. By the time it gets good and dark you will be glad you have some warm clothes to slip into. Over the years many folks have stopped observing while under a sparkling clear sky because they were too cold to continue (See Figs. 4.3 and 4.4).

Because your hands are exposed and are being used to change an eyepiece or take notes, the hands are very susceptible to cold weather. Therefore, gloves are needed, but they cannot be too thick or you can't do delicate tasks. Try a pair of rock-climbers' gloves. The ends of the fingers are cut out and the ends stitched to stop fraying. They allow the wearer to have the fingertips available and yet keep your hands quite warm. I also use a chemical hand warmer. They are a packet of chemicals that will warm up when exposed to the air. Once started up I find that they last for hours and keep me comfortable on a chilly night.

Put on the warm clothes *before* you get really cold. Trying to warm up after letting yourself get cold in difficult. The best technique is to wear several layers. The first layer is a light jacket or sweater. The next layer is the ski suit or heavy parka and muffler. The last layer is the cab of the truck or RV with the heater on full steam! I usually employ the last layer while driving home and complaining about how cold it is outside.

Take a Break

Try and take a break every couple of hours and get off your feet. A.J. likes to put his feet up on a chair, to help relieve the strain of several hours standing at the eyepiece. If you can be seated while observing, that is a more relaxing setup. During the break you can chat, look for meteors, or try and find an obscure constellation. A snack helps keep you awake and will give you energy to fight off cold. (No, it doesn't need to be a Milky Way bar!)

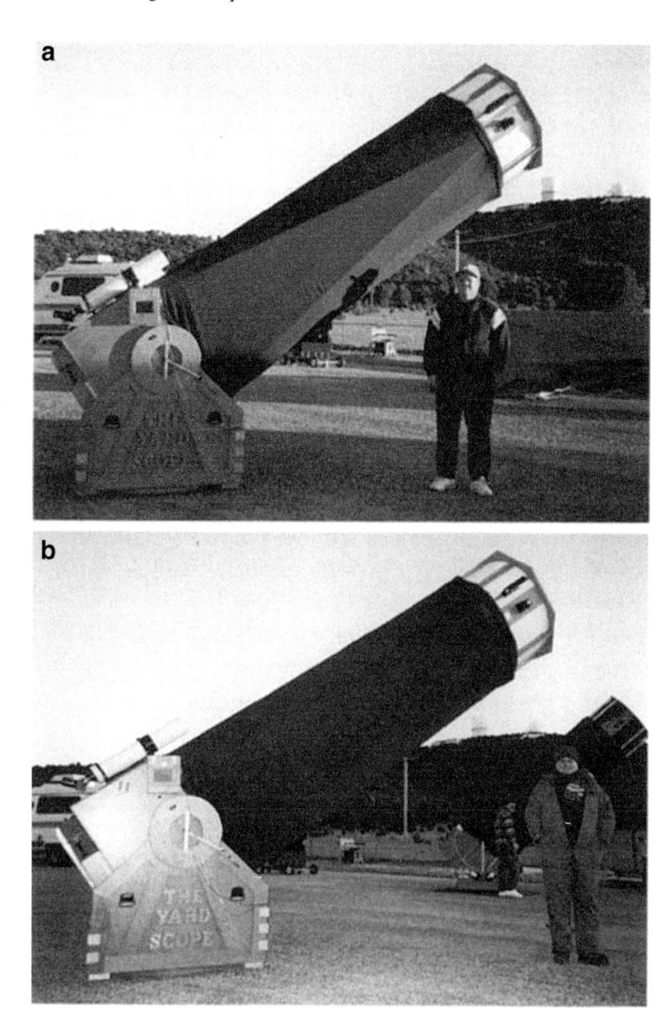

Fig. 4.4 (**a**) Before it gets cold. (**b**) All bundled up and ready for a night with a BIG telescope

A Nap Is a Good Observing Technique

It certainly was easier to stay up all night when I was younger. Getting a nap in the afternoon before an observing session makes a tremendous difference to how much observing will get completed. Another technique is to lie down for an hour or so at midnight, when some objects you wish to observe are on the eastern horizon. After a short rest, you will be ready to hop up and start observing. Either method will work, but get some rest. A really sleepy observer is not having fun and is not making good observations.

Make a Good Observing List

If you take the time to create a list of the deep-sky goodies you want to observe, it will make for a much more enjoyable evening. It is certainly fun to just put in a wide-angle eyepiece and sling the telescope from object to object. But pretty soon that method starts to be repetitive and you don't add any new objects to your notebook. A conscious effort to create a list of objects that you plan to look at will add much to your observing evenings.

There are plenty of sources for new things to observe. This is part of the good news for the modern observer. Between all the source books (several of my favorites are detailed in Chap. 18) and the wealth of information to be found on the Internet, don't be concerned about running out of new deep-sky wonders to see.

Recently, I took on a project to re-observe many of my old favorites. This turned out to be an excellent observing list generator. It had been many years since I had spent some time with the brightest deep-sky objects. I looked for new details or just made certain that I had a good observation of these bright galaxies, clusters and nebulae. So, you don't need to make that observing list just little, faint fuzzies; it can include bright objects that you just haven't observed in quite a while.

Get There Early

Arrive *before* sunset or just as the disk of the Sun disappears below the horizon. This gives you over an hour of twilight to set up and get ready to observe. There are two very good reasons to set up early. First, it gives the telescope some time to "air out" before being used. The mirrors can cool down to the temperature of the surrounding air and stop generating air currents inside the tube which degrade the image. This is particularly important during an Arizona summer (See Fig. 4.5).

The second reason to arrive early in the evening is that you can avoid using white light to get set up and therefore people will not throw rocks at you. If you are alone (probably because you turned on a white light flashlight last time) then not using the white light will allow your eyes to get dark-adapted sooner. That way, you can be ready to observe by the end of twilight.

Make Certain That Your Telescope Is Ready to Perform Well

After you acquire some observing skills you will expect your telescope and eyepieces to deliver excellent images. That can only happen if you take very good care of it. Make certain that the optics are clean.

If you have a commercially made telescope, follow the manufacturer's instructions about cleaning (or not cleaning) the optical surfaces. Even quite a lot of dust on the optical surfaces won't hurt the resolution or light-gathering power of your telescope, but damage to the optics caused by clumsy or inappropriate cleaning will.

Fig. 4.5 Setting up for a night of observing

I carefully wash the mirror in my Newtonian about once a year. Remove the mirror from the scope, blow off any grit with compressed air in a can and then let it soak for 10 min or so in distilled water with a drop of dish soap to loosen particles. Now use cotton balls with no finger pressure to wipe off the mirror. Drain the water off and tilt the mirror up, pour a solution of five parts of distilled water to 1 of isopropyl alcohol across the face of the mirror until it sheets off and leaves the mirror almost dry. If the water and alcohol don't sheet off the mirror, then start the process again. Keep going until the mirror is so clean that almost all the solution drains away. If there are a few drops left, remove them with a small piece of paper towel, only touching the water drop, not the mirror surface.

This cleaning is only needed once a year in a dusty desert because I keep the telescope covered well when it is not in use. Use a plastic sheet, mirror cover or even a trash bag to keep the dust and dirt from settling onto the scope in the first place. A.J. uses two plastic shower caps at the ends of the tube on his 8 in. f/6. An old film canister will fit snuggly into a 1.25 in. focuser opening. Seal up the scope against dust and it will last much longer.

Have a Telescope That Is Easy to Set Up and Use

I dearly love *large* telescopes, the bigger the better. But, they do have the huge disadvantage that they are—*huge*! It takes time and effort to get a big scope ready to use. That means that on lots of nights you will either set up the big scope and just grin and bear it or not observe.

Fig. 4.6 The 6 in. f/6, an easy-to-use telescope

For over 14 years my large telescope was a 13 in. f/5.6 Newtonian on a German equatorial mount. It takes about half an hour or so to get it ready to observe. In the heat of an Arizona twilight that is no fun. It is, however, a great cardiovascular workout.

This is the reason that I really enjoy my 6 in. f/6 on a simple alt-az mount. It can be set up and ready to use in less than 5 min. Great for a short observing session in the city and also fun on a night out of town when I just don't feel like hefting the big scope (See Fig. 4.6).

I have learned another thing since owning a smaller telescope. They provide terrific views of large deep-sky objects. The view with a rich field telescope of a wide variety of Milky Way star clouds and nebulae is stunning. In the large scope something like the North American Nebula in Cygnus would never fit in the field of view. However, in the 6 in. at 25× the entire shape of the nebula can be seen and appreciated. It also provides stunning views of the Milky Way star clouds: Scutum, Cygnus, Sagittarius, Norma, Crux and Carina. Just put in an eyepiece of between 40 and 20 mm focal length and start scanning. You will see a myriad of lovely curved star chains, small clumps of stars, dark rifts among the dense stars and a variety of star colors (See Fig. 4.7).

Fig. 4.7 A rich field telescope (RFT) will really show the Milky Way at its best

Much has been written and discussed about specialty telescopes for high magnification. The Moon, planets and close doubles demand precision optics and long focal lengths. On the other hand, much less has been discussed about the low-power end of the magnification spectrum. There are limits; in the next chapter I will talk about exit pupil size and how it restricts the lowest power a telescope can provide. Stay within those limits and enjoy beautiful views of a wide swath of the sky.

To put it simply, for easy, fun observing a small telescope can be a joy to use. Apertures between 3 and 6 in. (75 and 150 mm) in focal ratios less than f/8 will show off wide-field vistas in all their glory. I have had really fine views with both reflector and refractor rich-field telescopes.

The reason there is no such thing as a general-purpose telescope is that we don't live in a general-purpose Universe.

Test New Equipment at Home Before Taking It Out of Town

If you have brand-new equipment to use, then try it out in your backyard first. Going all the way out to a dark-sky site and then finding that a new bracket does not quite fit onto the scope or a new drive gear needs more lubricant to work smoothly wastes that precious dark-sky time. I learned this lesson when I borrowed a scope for a weekend. David Fredericksen was kind enough to loan me his 12.5 in. f/6 because my own 13 in. Newtonian was still in the process of being built.

Only when I had traveled far out of town and set everything up did I realize that this telescope was somewhat taller than me. Embarrassing. I should have brought along a small stepladder, something a backyard setup session would have made clear.

Fortunately A.J. was driving his trusty ’66 Chevy at the time and the spare tire was easily available in the trunk, so I borrowed it to use as a step. (Please notice at this time that I have borrowed all my essential observing equipment: the telescope and the spare tire “ladder”. It’s great to have astronomical friends who will put up with a mooch like me.)

Let’s see, where were we? Yes, it is a quiet midnight and I am standing on a tire intently observing a lovely planetary nebula in Cygnus. Without realizing it, I step on the air valve. Have you ever heard a rattlesnake? The buzz of a rattlesnake is amazingly similar to the hissing sound made by stepping on the air valve of a tire. Without thinking I leapt into the air, simultaneously breaking both the Olympic high jump and long jump records (but fortunately not the telescope). It took at least half an hour for my heart rate to return to normal.

So try out that new mount, focuser, camera or entire telescope before you leave on a long trip away from home. Remember the Boy Scouts’ motto: Be Prepared!

Chapter 5

What Other Accessories Are Useful?

Flashlights in the Dark

If the purpose of a trip away from city lights is to find lots of deep-sky treasures and then see detail within them, your eyes must be dark-adapted and stay that way. Using a white-light flashlight will prevent that process. Make certain that you use only red light to read star charts and write notes. I have found that it is easier to read by the light of a red light-emitting diode (LED). The pure red light doesn't seem to affect my night vision as much as a white flashlight with a red filter. Several manufacturers of astronomical equipment now have red LED flashlights available. They are somewhat expensive for what you get (an LED costs a few cents), but worth it nonetheless (See Fig. 5.1).

Bring an extra red flashlight and some batteries. Few things can stop a great night with the telescope faster than a dead flashlight. You can put a piece of masking tape over the end of the light to diffuse the glow when the batteries are new, then remove it as the batteries weaken. Use a lanyard so that the flashlight can be hung around the neck and so can't be misplaced—what are you going to use in the dark to find a missing flashlight?

The "Monk's Hood"

While going after the very faintest objects even the dim light which surrounds you at the eyepiece can affect your vision. To block off that extraneous light a dark cloth which covers the head is very useful. A.J. and I discovered this fact for ourselves

S.R. Coe, *Deep Sky Observing*, The Patrick Moore Practical Astronomy Series,
DOI 10.1007/978-3-319-22530-2_5

Fig. 5.1 A red flashlight will allow you to read star charts and notes without affecting your night vision

years ago and have found it to really focus our attention on faint details in the field of view. Any type of dark cloth will do, as long as it is large enough to cover your peripheral vision and make the field of view at the eyepiece the only light you are seeing (See Fig. 5.2).

A rubber eyecup on your eyepieces is somewhat helpful in blocking out unwanted light from getting into your eye as you observe. However, I don't find an eyecup as useful as a dark cloth covering my entire head.

It seems that two effects are happening when I use the "monk's hood". First, the fact that some external light is being removed means that I get better dark-adapted. Second, the concentrating effect of only seeing the light that is coming out of the eyepiece means your entire visual mind is centered on just that information. It is certainly the least expensive accessory in this book. I bought a yard square (a square meter) of dark-blue cloth; A.J. uses a black towel.

Tools for Collimation

Nothing you do will improve the performance of your telescope more than making certain it is in good collimation. This means having the mirrors correctly aligned so that the image is not distorted by a poorly placed mirror. This is not an easy skill,

Fig. 5.2 Pierre and A.J. use a hood during the daylight to video sunspots

and if there are some knowledgeable people who can help you the first time you try collimating your scope, be nice to them. (Astronomers like pizza.)

This discussion concerns two types of telescopes: Schmidt–Cassegrains and Newtonians. Refractors have their lenses adjusted at the factory and need little further adjustment unless they are damaged. Maksutov telescopes also need no user adjustment. The mirror corrections in Schmidt–Cassegrain telescopes (SCTs) and Newtonians are generally made by adjusting three screws that are 120° apart. They tilt the mirror so that, when the system is correctly collimated, an out-of-focus star consists of several perfectly concentric circles. Don't forget to take the screwdriver or Allen wrench that fits your telescope.

Several companies provide collimation kits. They include the collimation tools and a booklet on how to use them. Also, C.R. Kitchin has a drawing of what a well-collimated system looks like in his book *Telescopes and Techniques* (Springer, ISBN number 0-387-19898-9). The focused star will be just a tiny dot.

The collimation of even the best-made reflecting telescope needs to be checked before use every time it is transported. It doesn't take much jostling to get those mirrors out of alignment. Arriving at an observing site well before dark will ensure that you have time to collimate your optics. It is not an easy task to deal with a mirror system that is tilted when you are in the dark: trying to shine a red flashlight down the tube and adjust the mirror cell at the same time is a job for an intelligent octopus.

Fig. 5.3 Use the collimator tools to align the mirrors in your telescope for best performance

A Guide to Eyepieces

One of the most confusing aspects of getting started in amateur astronomy is the wide variety of eyepieces available. If you get a group of observers together and let them chat for long enough, the talk will turn to eyepieces sooner or later. The reason is that eyepieces are the most important accessory you can purchase for your telescope. Once you have chosen a telescope, then all the characteristics of that astronomical viewing system are determined by the eyepiece. I always think of an eyepiece as a small magnifying microscope which allows me to inspect the image formed by the mirror in my telescope. If you aim your scope at the Moon, then your optical system will create a tiny image of the Moon in midair: your eyepieces let you observe that image with varying magnifications and fields of view, according to the eyepiece used.

Let us begin by getting some terminology straight. Here is a list of the definitions of some words used in connection with eyepieces.

Apparent Field of View

This is the width, in degrees, of the field as seen through just the eyepiece alone. If I have two eyepieces with the same focal length then the one with the larger apparent field of view will show more of the sky if inserted into the same telescope.

This parameter is determined by the design of the lenses inside an eyepiece. The older Plossl design has an approximately 50° apparent field, while many modern designs have an approximately 80° apparent field. So if you point your telescope at a starry Milky Way field, the 80° eyepiece will provide a much wider view.

Curvature of Field

Good eyepieces provide a field of view which is flat. The focused image should be sharp from edge to edge. Star fields are a tough test of this characteristic.

Distortion

Good eyepieces also have little distortion, which means that if you viewed a piece of lined graph paper all the lines would be straight and would cross at right angles. Distortion can be a problem for only a small section of the field of view, but curvature generally happens to the entire field of an eyepiece.

Exit Pupil

The lenses in an eyepiece, in conjunction with the lens of your eye, focus the image on your retina. When observing, you position your eye so that it can see this image at its brightest. The diameter of the "bundle" of light rays entering your eye obviously has to be small enough to go through your eye's pupil—about 1/3 in. (7 mm) at most. If everything is working as it should, the light rays will all pass through your pupil. If the bundle of rays is too wide, light that would otherwise form an image on your retina is wasted because it cannot fit into your eyeball. As I am writing this second edition, I am 66 years old and have been viewing the sky for 40 years. My elderly eyes just don't open up like they used to, in my 30s I could see 14 Pleiades, now 9 can be seen on an excellent night.

Eye Relief (See Fig. 5.4)

This is the distance from the eyepiece lens to your eyeball. This value is important to eyeglass wearers. If you need to have your glasses on to view the sky, there must be plenty of eye relief so that your eyeglasses will fit between the eyepiece and your face. Those of us who don't wear glasses to observe generally like some eye relief to avoid the feeling that the lens of the eye is being jammed against the glass lens of the eyepiece.

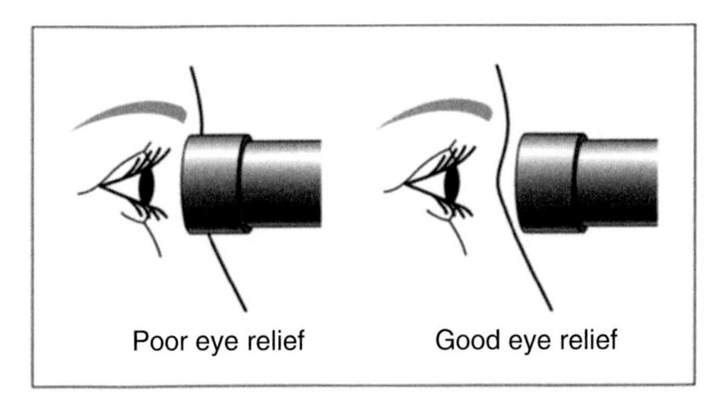

Fig. 5.4 Good eye relief provides comfortable viewing

Focal Length

The apparent distance from the lens to the object being viewed, in this case the image formed by your telescope. Long-focal-length eyepieces show a large portion of the image being viewed and short-focal-length eyepieces will allow a small section of the image to be inspected at higher magnification. This is how you choose the magnification of your optical system. Pick out a long-focal-length eyepiece, say 40–24 mm, and the system will give a wide field and low power. Select a short-focal-length eyepiece, around 8–4 mm, and you will get a high-power, small-field-of-view look at whatever is in the scope. The magnification of any telescope is calculated by dividing the effective focal length of the primary mirror (or optical system) by the focal length of the eyepiece. Thus a Newtonian with a 2 m focal length, fitted with a 24 mm eyepiece will give a magnification of about 83 times.

Ghost Images

With a poorly made eyepiece some of the light from a bright star can reflect about inside and form faint images within the field of view. These ghost images can be subdued by multi-coating the lenses in the eyepiece, and by proper design of the eyepiece optics. Only the cheapest eyepieces nowadays are not coated.

Real (or "True") Field of View

This is the field of view of the entire telescope system, including the eyepiece. It varies with the magnification you are using.

The previous information dealt with some key phrases, so let's move gently into a little calculation concerning eyepieces. There are four formulae that apply to using and understanding the values associated with eyepieces. These are:

$$\text{Telescope focal length} = \text{Focal ratio} \times \text{Primary mirror}\left(\text{or lens}\right)\text{diameter}$$

$$\text{Magnification} = \text{Telescope focal length} / \text{Eyepiece focal length}$$

$$\text{Exit pupil} = \text{Telescope aperture} / \text{Magnification}$$

or

$$\text{Exit pupil} = \frac{\text{eyepiece f.l.}}{\text{f-ratio}}$$

$$\text{True field of view} = \text{Apparent field of view} / \text{Magnification}$$

Just remember that 1 in. equals roughly 25 mm and you are ready to figure out these values for your telescope. Grab your calculator and try a worked example of some scope and eyepiece combinations. Assume you have a 6 in. f/8 telescope. That means the scope has a 48 in. focal length (from the first formula above, 6 in. times f/8). Converting 48 in. to mm (because for some reason eyepieces are always measured in millimeters) gives 1200 mm (48 × 25).

Let's say you have three eyepieces which have focal lengths of 20, 12 and 7 mm; here are the magnifications each will provide:

60× for the 20 mm eyepiece (from 1200 mm/20 mm);
100× for the 12 mm eyepiece (from 1200 mm/12 mm);
171× for the 7 mm eyepiece (from 1200 mm/7 mm).

Now, here are the exit pupils for those eyepieces (remember, you had to convert 6 in. of aperture to 150 mm first):

2.5 mm exit pupil for the 20 mm eyepiece (from 150 mm/60×);
1.5 mm exit pupil for the 12 mm eyepiece (from 150 mm/100×);
0.88 mm exit pupil for the 7 mm eyepiece (from 150 mm/171×).

To figure out the *real field of view* for each eyepiece, we need to know the apparent field of view for the type of eyepiece used. Let's assume you are evaluating eyepieces with an apparent field of 60°—this is a good average for a middle-of-the-range commercial eyepiece.

1° FOV for the 20 mm eyepiece (from 60°/60×);
0.6° FOV for the 12 mm eyepiece (from 60°/100×);
0.35° FOV for the 7 mm eyepiece (from 60°/171×).

Because the true FOV is often less than 1°, this value is generally given in arc minutes. There are 60 arcmin in 1°. So, 0.6° × 60 equals 36 arcmin as the true FOV of the 12 mm eyepiece. Also, 0.35 × 60 means that the 7 mm eyepiece provides a 21 arcmin field.

I know that all this math is not particularly fun, but it does give some useful results. We can draw some general conclusions from our results. As the power is increased in your telescope, you get a smaller exit pupil and a narrower field of view. Because the pupil of your eye cannot generally get wider than 1/3 in. (7 mm), it is not useful to buy an eyepiece that gives a larger exit pupil than that. The news is worse for those of us in advanced puberty; if you are over 45 years old, your eye probably does not open larger than 5 mm. At the other end of the scale, magnifications which yield an exit pupil smaller than 0.5 mm are not very useful either. It turns out that your eye has its best resolution if provided with an exit pupil of about 2 mm.

Every set of your eyepieces should provide a magnification that gets the system close to this optimum resolution value.

To calculate the widest field of view that a telescope can provide just remember the formula:

$$\text{Exit pupil} = \text{Telescope aperture} / \text{Magnification}$$

Now rearrange the formula (you thought there wouldn't be a math quiz?):

$$\text{Magnification} = \text{Telescope aperture} / \text{Exit pupil}$$

So if we stick with our 6 in. (150 mm) scope and plan to have it give an exit pupil of 5 mm (remember the upper limit even for a teenager is about 7 mm), then 150 mm/5 mm gives 30×, which is the *lowest* limiting magnification this telescope can give without exceeding the widest exit pupil we have chosen. Pick an eyepiece which will give 30× and you are at the widest field of view this telescope can yield.

If you are looking for great, wide-angle views of the Pleiades or a Magellanic Cloud, then get an eyepiece with a long focal length and a wide apparent field of view. Are you looking for Encke's division in Saturn's rings? Or maybe trying to spot detail within planetary nebulae? Then look for an eyepiece with a short focal length and still with good eye relief so you don't have to jam your eye into the glass to observe. Maybe you wish to observe star clusters and galaxies at their best? Then get an eyepiece that provides that magic 2 mm exit pupil.

In the same way that there are several different designs of telescope optics, the lenses within an eyepiece are arranged in a variety of ways to provide the observer with magnification and a focused field of view to observe. Basically, as lens makers got more consistent results, they realized that adding more elements (lenses) could reduce the aberrations in the eyepiece. Therefore a lot of the history of eyepiece design is a study in adding more lens elements to each new eyepiece. Many of the designs I will discuss are named after the person who invented the combinations of lenses which make up these eyepieces. This is not just a history lesson, however; all these eyepieces are available on the market today.

The first eyepiece designs, the *Ramsden* and *Huygenian,* only contain two lenses and are very poor performers by modern standards. They have very narrow fields of view, short eye relief and many aberrations. Cheap telescopes often include these inexpensive eyepieces.

The *Kellner* is the best of the inexpensive eyepieces. It has been around for many years and contains one doublet (two lenses together) and one singlet lens for a total of three pieces of glass inside. The Kellner does not have any excellent characteristics,

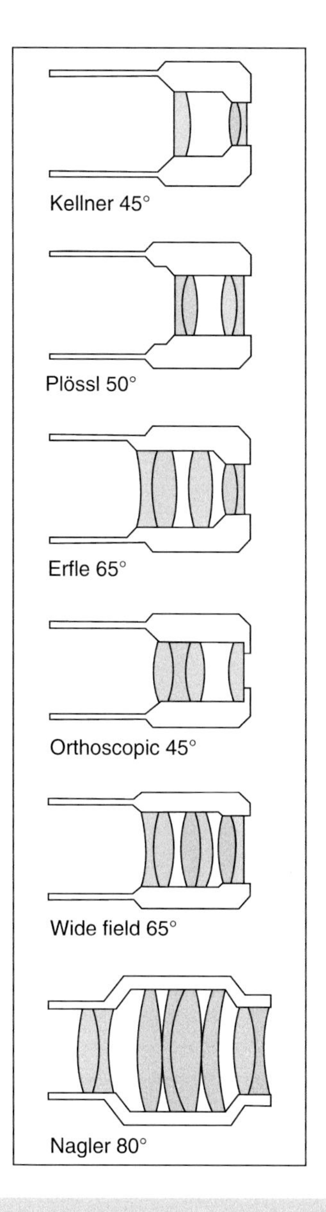

Fig. 5.5 Eyepiece types

but it also has few real flaws. Kellner eyepieces have mediocre eye relief, a fair field of view (45°) and little curvature of field (See Fig. 5.5).

The *Plössl* eyepiece is composed of two doublets, which are identical to each other. For this reason, you will also hear it called a symmetrical eyepiece. It is an excellent eyepiece and many observers look no farther than a good set of Plössls. They have a fairly wide field of view (55°), good eye relief and are well corrected for aberrations. They cost more than Kellners, but they are worth it.

Orthoscopic eyepieces are generally not named for their inventors, Mittenzwey and Abbe, and I think you can see why. The "Orthos" have one outstanding characteristic, namely that the aberrations and distortions in these eyepieces are very well corrected. These flat-field eyepieces have a modest amount of eye relief and field of view (50°). This design contains a triplet lens with one singlet nearest your eye.

The *Erfle* eyepiece was invented to provide a wide apparent field of view, and they do that (65°). What the Erfle design gives up is some sharpness of the image at the edge of the field. Also, if there is a very bright star near where you are observing, some ghost images can appear within that wide field. Inside the Erfle is a combination of two doublet lenses and one single.

This is where the eyepiece world stood for many years. All these older designs are best when used in telescopes with f/ratios longer than f/6. Then the advent of computerized lens designs changed the standards for eyepiece manufacturers.

Enter the *Nagler* and *Ultra Wide* designs. These computer-designed eyepieces contain either seven or eight lenses, some with curves ground into them which would have been impossible before modern grinding machines were constructed. These designs provide an extremely wide field of view (82°) and very distortion-free fields at those wide angles. They all have two disadvantages: cost and weight. All that glass is going to cost more to grind and put together, also, once it is assembled these eyepieces weigh nearly 2 pounds (almost a kilogram) in long focal lengths.

In the time between the two editions of this book, several manufacturers of eyepieces have gone even father to include eyepieces with fields of view of 100°. They are the largest, heaviest and most expensive eyepieces now on the market.

Now, you are going to get an opinion from me and it is nothing else. I don't see enough difference between 80° and 100° to make it worth the money. These extremely wide eyepieces force you to swivel your head to see the entire field and I don't wish to do that for all that cash. I will provide you a saying from the members of the Saguaro Astronomy Club, "don't look through more eyepiece than you can afford".

Along with eyepieces themselves, there is a device which will change the magnification of your system. It is called the *Barlow lens*. Just slide your eyepiece into the Barlow and put the whole thing into the eyepiece focuser and you have raised the magnification. The good news is that the eye relief of the system is the eye relief of the eyepiece alone. So for high power it is much easier to use a 10 mm eyepiece and a 2× Barlow than it is to use a 5 mm eyepiece and its short eye relief.

This is a great idea and I have owned a Barlow since my first scope, but there are limits. Barlows that provide more than 2× are also introducing so much magnification to the viewing system that I believe I am seeing more detail than I saw without the Barlow in place. So, use your Barlow in moderation and purchase a Barlow that magnifies either 1.8× or 2× and it will prove to be a very useful device.

Now that you know enough (for now) about eyepieces, I'll bet you are still left with the same question: "Which ones do I buy?" (See Fig. 5.6)

That is a tough query, but I will try to help. If you are just getting started, purchase three eyepieces and a Barlow lens in the beginning. Buy one low-power, wide-field eyepiece which has a focal length between 25 and 20 mm. Get one

Fig. 5.6 Several different eyepieces are needed to give different magnifications and fields of view

medium-power eyepiece, from 16 to 12 mm. Buy one high-power eyepiece, from 9 to 6 mm focal length. Get a Barlow lens with a magnification of either 1.8× or 2×. With those four things you will be prepared to observe a wide variety of what there is to see in the sky.

As time goes by you can fill in as much as your budget will allow. You might choose a really wide-field 32–35 mm eyepiece. Or maybe something between the medium and the high power. I know that if you are just getting started, you might be thinking about a very high-power eyepiece of about 4 mm focal length. Even though it seems nifty to have a scope that can go to 600×, the number of evenings steady enough to use extreme magnifications is rare. You can make use of very high power occasionally, but not often.

Eyepiece design is the subject of much talk whenever astronomers compare eyepieces and figure how much money they have to spend. If you can afford it, at least start with a medium-power Plossl, a high-power Orthoscopic, and a wide-field Erfle. If you are really in a pinch for money then Kellners will suffice. However, if you go out and observe with other folks who have better eyepieces than yours, it can be an expensive trip. One spectacular view through someone's brand-new pride-and-joy eyepiece can have you looking through catalogs and checking the limit on your credit card.

I have had a variety of different eyepieces over the years. My first scope was an 8 in. f/6 and it had a 1.25 in. focuser, so all my eyepieces were that size. I used three Erfles: 20, 16 and 12 mm for medium-power viewing. When I first got the scope, I did what I have told you *not* to do—I ordered it with a 4 mm eyepiece and never saw a clear view in it. I was able to trade the 4 mm for a 6 mm Orthoscopic that became a prized eyepiece for looking at fine detail on the Moon and the planets. Once I added a 2× Barlow, I was set and my eyepiece collection changed little for several years.

When I sold the 8 in. to finance a 17.5 inch Dobsonian (yes, aperture fever got me too), I decided I needed a 2 in. focuser and an eyepiece to fit.

Luckily, I found a war-surplus 38 mm Erfle that only needed some machining to make a sleeve that would fit the 2 in. focuser. A friend with a lathe made the part and I was in business. Again, my eyepiece collection seemed complete for a while.

In the 1980s the Nagler revolution hit. The first 13 mm Nagler eyepieces I used had a problem that was serious for some observers, including me. Some folks see a "kidney bean", a dark marking within the field of view, which will not go away regardless of how the observer moves their eye or head. I did not view this as a problem, because it prevented me from spending the money for these expensive eyepieces.

However, Meade decided to release its Ultra Wide series and I got a chance to use the 14 mm at the Riverside Telescope Maker's Conference. That was the final straw. The wide field of view, generous eye relief and excellent contrast of these eyepieces sold me. I found someone to buy my old eyepieces and I have recently completed the set of Ultra Wide eyepieces. In the same time, I also used and then bought a 22 mm Panoptic eyepiece that is excellent in my 13 in. f/5.6 Newtonian. The wide, flat, contrasty field of view of the *Panoptic* is stunning and it has become one of my favorite eyepieces. Now that I know how good the Panoptic design is, I plan to upgrade my widest-field eyepiece to a 35 mm Panoptic. My plan is to use some of the money from this book to purchase my 35 mm Panoptic, so thank you for contributing. All the previous conversation is from years ago and I have owned and used that 35 mm Panoptic very often, still to this day.

I have two types of eyepieces. My Panoptics are 65° eyepieces and in a 2 in. barrel they provide excellent wide, sharp and flat views of the sky. I own the 35, 27 and 22 mm. I also have three of the Meade Ultra Wide Angle eyepieces. They have an 84° field. I own the 14, 8.8 and 6.7 mm. They provide excellent views at a variety of higher magnifications. I bought them at the Riverside conference many years ago and have never looked through an eyepiece that made me want to trade them in for something newer. Eyepieces are a personal choice, so try and view through an eyepiece that you are considering before buying it. Or, make certain that you are dealing with a company that has a good return policy.

Here is a place where being a member of an astronomy club is worth the time, money and effort. If you are going to go out and observe with other club members you will get an opportunity to view through a variety of eyepieces, telescopes, finder scopes and other equipment. The only way I can give the opinion in the previous several paragraphs is that I have viewed through lots of eyepieces over the years. If you do join a club or go to a large star party, make time to walk around and view with other people's equipment.

Once you have purchased the eyepieces that will serve you, make certain they are well cared for. A good eyepiece box is a small investment when you add up the price of several good pieces of glass. Be careful when cleaning your eyepieces. Never rub them with any real force, always gently, or you will scratch the coating. Use a squeeze bulb or canned air to blow off any dust before cleaning. I use a cleaning rag I purchased at a camera store which does an excellent job removing the greasy fingerprints which inevitably happen. Microfiber cloths intended for cleaning spectacles are excellent.

If you are thoughtful in purchasing good eyepieces and then protecting them from the elements; they will last many years and provide you with spectacular views of the heavens.

Light Pollution Reduction (LPR) Filters

Open any issue of a magazine about astronomy and there will be advertisements for filters. For deep sky, I have never found a useful purpose for color filters. Red, orange, yellow and blue filters do a wonderful job on the moon and planets, but that is for a different time and another book.

LPR filters will reduce the amount of light pollution in the field of view that is being visually observed or photographed. This means that the deep-sky object will stand out from the glowing background more easily. Because the Sun is yellow and city lighting engineers wish to provide street lights that mimic the Sun, most light pollution is bright in the yellow part of the spectrum.

Fortunately, most emission from nebulae is either greenish-blue or red. The LPR filter can let the light from the astronomical object through the filter and block off the light from street lights and football fields. In the chapters that are about planetary and emission nebulae, some of the objects were chosen to demonstrate the improvement that these filters provide (See Fig. 5.7).

There are three general types of LPR filter. The *wideband* or *broadband* filters pass most of the visible light and only block off a small portion of the spectrum. This allows the broadband filters to raise the contrast on long-exposure photographs, but they do little for visual observers. The Orion Skyglow, Lumicon Deep Sky and Meade #908B filters are of this type.

The *narrowband* filters will pass two or three narrow colors of visible light; they are manufactured to allow through the most prominent colors seen in emission and planetary nebulae. The Lumicon Ultra High Contrast (UHC), Meade #908N and Orion Ultrablock are such narrowband filters.

The third type of LPR filter is the *line filter.* This filters all of the colors except one. There are several of these very narrow or line filters, the most common being the Oxygen III. The O-III filter passes an oxygen emission line that is in the blue–green. This line is quite prominent in both gaseous and planetary nebulae. Both Lumicon and Meade have O-III filters in their catalogs.

The *H-Beta line filter* is even more specialized. It certainly works with the California Nebula in Perseus (NGC 1499) and also the Horsehead Nebula (IC434 and B33) in Orion. However, the largest list of deep-sky objects that I have seen which respond to the H-Beta filter is about 10, and the last 5 of these were very faint.

Before I spout my opinion about these filter types and what they can do for your observations, let me tell you that on this subject there is pretty heated discussion among observers. I believe this is because we are talking about human brain–eye interaction, and that can get personal. So, as A.J. says, "This is a religious discussion."

Fig. 5.7 A wide variety of filters will thread into an eyepiece

I have owned an LPR narrowband filter for 35 years, the Lumicon UHC. It has provided a lot of help in seeing detail within pretty bright nebulae and also detecting faint nebulae. On several occasions I have had a chance to use an Orion Ultrablock and have found them to be just as useful as the Lumicon. I saw no difference between the two filters I had a chance to test.

I have also had several chances to use a wideband filter, but I have never owned one. There is not enough difference between the view with the filter in place and that without the filter to justify the cost. There are plenty of deep-sky astrophotographers who use wideband designs and get great results on the photos. That seems to be the most useful function for the wideband design.

When it comes to the line filters, I disagree with many deep-sky observers. My eye doesn't see much difference between what a narrowband and a line filter can provide. I owned an O-III filter for almost a year and had many opportunities to use it in a test alongside the UHC filter. I never saw any detail with the O-III that I could not also see with the UHC filter. Also, I did not like the effect of the O-III on stars; it makes them reddish purple to my eye. The view is a little too psychedelic for my taste.

That is what needs to be emphasized here. This is a matter of taste. So, my advice is start with a narrowband filter for the best overall results. It will certainly raise the contrast and detail in a wide variety of nebulae.

One important point to remember is that modern filters use something called the "Fabry–Perot" effect that can achieve a very narrow passband, which is why they can be very effective at filtering out sodium or mercury street lights. The main adverse side-effect is that any LPR filter will result in at least a one-magnitude loss of star brightness for visual observers.

Chapter 6

Why Should I Take Notes While Observing?

My journey toward becoming a more experienced observer and really *seeing* the sky with all its charms can be traced to the time when I started taking good notes. My first observing notebook is filled with cryptic things like "Wow" or "Not much". You can do better than that! If you take the time to find it, at least document it with some description of the object. Even if you don't plan to ever look at this faint fuzzy again, give it a decent epitaph.

My astronomical hero is Sir William Herschel. I won't veer off into a history lesson, but if you are interested, his was a fascinating life. Suffice it to say that during the early nineteenth century he discovered many deep-sky objects using telescopes he made with his own hands. Herschel said "seeing is an art that is learned".

Taking consistent and detailed notes will provide excellent practise at the art of seeing. It is all too true that you can't get experience by wishing for it. Enjoy the journey toward becoming an experienced observer and the Universe can become a fascinating path to walk, rather than a task to be completed.

Unfortunately, modern astrophotography does confuse people seeing their first view of a particular deep-sky object. They point to a colorful, star-filled picture with glowing arches of nebulosity and ask, "Why doesn't it look like that?". The reason is that modern cameras can gather light for a much longer time than your eye. Therefore a long-exposure photograph shows much more detail than can be visually seen with a telescope.

I realize that a great picture is a real treasure and a way to capture the beauty of the sky. But astrophotography is not the only way to acquire and keep your favorite deep-sky vista. Taking detailed notes and making a good drawing will also let you hold the Universe in your hand.

S.R. Coe, *Deep Sky Observing*, The Patrick Moore Practical Astronomy Series,
DOI 10.1007/978-3-319-22530-2_6

I like looking at a great picture, but there is no substitute for standing in the presence of the original.

Some years ago had the pleasure of a vacation in New York City. A day at the Metropolitan Museum of Art will provide you the ability to stand in front of original paintings by a fascinating variety of artists. Owning a book with images of a painting by Rembrandt or Van Gogh is definitely *not* the same experience as standing before the original. The same is true of observing the heavens; no photograph of the Orion Nebula is the same experience as seeing it for yourself. That light is at the end of a journey of 1500 years that culminates within your eye.

Another point to be made is that a media system which thrives on its ability to sell people everything from beer to automobiles has deadened your senses. As radio, television and your computer bombard you with images of everything from dancing girls on the hood (okay, "bonnet" in the UK) of a car to talking frogs; your mind has learned to shut it out.

You have to make a conscious effort to open your eyes and use all the information they are gathering. We live in a flashy world and you need to get beyond the flash to some real substance. Noting what you observe will help.

What Type of Notes Should I Take?

For an answer to "What do I say?" we need look no farther than Sir William Herschel. As the most prolific and famous deep-sky observer of all time, he invented a method of shorthand for observing notes that give lots of important information about what is seen in the eyepiece. Herschel's notes give information on total brightness, size, shape, brightness contour, resolution into stars, unusual features and nearby stars. His work, and his son John's, are the basis of the New General Catalog, the NGC. So, here are the abbreviations used by the NGC:

!	remarkable object
!!	very remarkable object
11m	11th mag
8…	8th mag and fainter
9…13	9th–13th mag
am	among
att	attached
B	bright
b	brighter
bet	between
C	compressed
c	considerably
Cl	cluster
D	double
def	defined
deg	degrees
diam	diameter

(continued)

(continued)

dif	diffuse
E	elongated
e	extremely
er	easily resolved
F	faint
f	following (to the E)
g	gradually
iF	irregular figure
inv	involved
irr	irregular
L	large
l	little
M	middle
m	much
mag	magnitude
N	nucleus
n	north
neb	nebula, nebulosity
nf	north following (NE)
np	north preceding (NW)
p	poor
P w	paired with
p	preceding (to the west)
p	pretty (before F, B, L or S)
R	round
Ri	rich
r	not well resolved, mottled
rr	partially resolved
rrr	well resolved
S	small
s	south
s	suddenly
sc	scattered
sf	south following (SE)
sp	south preceding (SW)
st	star or stellar
susp	suspected
v	very
var	variable

If you have never dealt with the NGC abbreviations before, perhaps a few examples will help:

NGC#	Description	Decoded descriptions
214	pF, pS, lE, gvlbM	pretty faint, pretty small, little elongated, gradually very little brighter in the middle
708	vF, vS, R	very faint, very small, round
2099	! B, vRi, mC	remarkable object, bright, very rich, much compressed

(continued)

(continued)

NGC#	Description	Decoded descriptions
4866	B, pL., mE, sbMN	bright, pretty large, much elongated, suddenly brighter middle with nucleus
5694	cB, cS, R, psbM, r, ★9.5 sp	considerably bright, considerably small, round, pretty suddenly brighter middle not very well resolved or mottled, star of 9.5 magnitude is south preceding
6643	pB, pL, E 50, 2 st P	pretty bright, pretty large, elongated at an angle of 50°, two stars preceding
7009	!, vB, S	remarkable object, very bright, small
7089	!! B, vL., mbM, rrr, stars mags13…	extremely remarkable object, bright, very large, much brighter middle, well resolved, stars 13th mag and dimmer

The most available reference source which includes the NGC descriptions is *Burnham's Celestial Handbook* by Robert Burnham, Jr. I can highly recommend this fascinating work. It includes lots of great information about each constellation and a ready-to-use observing list of the best objects in that part of the sky.

There is another book which will provide you with lots of objects of observe. That is the *Night Sky Observers Guide*. It contains observations, drawings and photos of plenty of deep sky objects.

You will be reading my observations of a variety of deep-sky objects later in this book. Examining the wording used by other observers will provide you with a vocabulary so that you can write and speak good descriptions. This isn't stealing—it is research.

A small tape-recorder is a help to many people, but make certain you transcribe your observations before you use that tape again. If you take written notes as I do, don't make the mistake of trying to use a red pen. The red ink will disappear under the light of the red flashlight.

I spend much of my time on the Internet at the Cloudynights web site. We discuss what we have seen and how to get better views of star clusters, galaxies, nebulae and comets. At least that is in the forums where I am reading and writing. If you are going to participate then you will need to take notes so that you can tell others what you observed. I have 100 articles that are in an archive on the Cloudynights site.

Rating the Night

One of the records you can keep which will provide useful information is a rating system for the night sky while you are observing. There are two pieces of data to record. One is the sky transparency, a measure of sky clarity or lack of clouds and haze. Because this rating will affect the ability to see faint objects against the background sky, it is often called "contrast".

The second value is "seeing", a measure of the sky steadiness or lack of twinkling. These two ratings seem to have little to do with one another. I have seen beautifully transparent nights when the stars twinkled on and off. One the other hand, there have been hazy nights when only the brightest constellations are seen, but the view at 300 × is steady and unwavering. I know that it seems unfortunate sometimes, but the column of air in front of your telescope is part of the optical light path, even if you don't want it to be. The reason the Hubble Space Telescope reveals those amazing images is that it is above the moving, hazy atmosphere.

Rating	What it means
0	Completely cloudy; no stars seen. (Why are you out?)
1	More than 50 % of the sky is cloudy.
2	More than 25 % of the sky is cloudy, less than 50 %.
3	More than 10 % of the sky is cloudy, less than 25 %.
4	No clouds, but hazy; only brightest stars seen down to 4th mag.
5	Somewhat hazy; some fainter stars seen, to mag 5; Milky Way visible only in brighter regions.
6	Not visibly hazy but Milky Way visible only in brighter regions (Sagittarius, Cygnus, Norma + Crux); stars seen to mag 5.8.
7	Fainter stars, equal to mag 6.0, are seen and the fainter parts of the Milky Way seen with averted vision; Zodiacal light seen with averted vision.
8	Stars fainter than mag 6.0 are just seen and fainter parts of the Milky Way are more obvious; Zodiacal light is seen with direct vision.
9	Stars fainter than mag 6.0 are seen with direct vision and so are faint portions of the Milky Way (Lyra, Libra); gegenschein seen with averted vision.
10	Overwhelming profusion of stars; Zodiacal light and the gegenschein form continuous band across the sky; the Milky Way is very wide and bright throughout.

This scale for transparency was invented by the members of *SAC* as we observed together. Generally, we agree on the rating for the night. Practice may not make perfect, but it does make for consistency.

I have rated a night as a "10" quite rarely. But, as you can imagine, those are very memorable observing nights. At the other end of the scale, I have observed on a one or two night when something special (occultation or eclipse) is happening, but it was a poor view of that object.

Next is a scale for seeing; it was originated by W.H. Pickering while he used a 5 in. (120 mm) refractor at Harvard Observatory. His descriptions of diffraction

Rating	What it means
1	Star image is usually about twice the diameter of the third diffraction ring if the ring can be seen; star image 13 arcsec can be abbreviated 13″ in diameter.
2	Image occasionally twice the diameter of the third ring (13″).
3	Image about the same diameter as the third ring (6.7″), and brighter at the center.
4	Central Airy diffraction disk often visible; arcs of diffraction rings sometimes seen on brighter stars.
5	Airy disk always visible; arcs frequently seen on brighter stars.
6	Airy disk always visible; short arcs constantly seen.

(continued)

(continued)

Rating	What it means
7	Disk sometimes sharply defined; diffraction rings seen as long arcs or complete circles.
8	Disk always sharply defined; rings seen as long arcs or complete circles, but always in motion.
9	The inner diffraction ring is stationary. Outer rings momentarily stationary.
10	The complete diffraction pattern is stationary.

rings and Airy disks can only really be applied to an instrument of that size. The Pickering scale works well for many people, however, and you can certainly modify it if you need to do so.

There are some similarities between these two scales. First, the larger numbers designate a better night to observe. Second, the values of the numbers also are similar: 1–3 is considered very bad, 4–5 poor, 6–7 good, and 8–10 excellent. I find this very helpful in keeping the ratings system straight. So, if you hear that a night was 4 out of 10 for seeing and a 5 for transparency, then you know it was a pretty mucky night with twinkling stars. On the other hand if the ratings are 7 for seeing and 8 for transparency that was a very good night at the telescope with clear and quite steady skies.

Astronomical Drawings

I can hear you saying "But I can't draw." Well, I can't either. If asked to draw a human being I would sketch a stick figure. But a drawing of an astronomical object is much easier—honest. What it can provide you with is a personal record of what you saw that night. A good sketch can provide detail that no written description can match. Don't show the finished results until you have practiced. No doubt Rembrandt had some paintings he kept in the cellar.

Before going out to the telescope, make a drawing form. Mine includes space for: date, constellation, telescope size, seeing, contrast, site and notes. Each object gets room for: object designation, filter used, size of field of view, magnification and notes. Then there is a circle in which to draw. Something around 3 in. (75 mm) across will do fine. If you are a computer user, then any good word processor will make this easily. If you are not a computer user, then a large can of soup can be used as a template for a drawing circle. Make a good printout of your form and check it for tiny ink blotches; they can be mistaken for stars.

Once out in the field a clipboard will prove very handy. I use only a #2 pencil (HB in the UK) when drawing at the telescope and then smear the graphite around with my finger for extended objects such as nebulae. On the dimmest objects, even the red flashlight on white paper will make those faint fuzzies disappear from the glare. Wait a few moments to reacquire your night vision and the object should

Fig. 6.1 Pencils and charcoal pencils, a blending stump, erasers and an eraser shield

reappear. Try and memorize the field and draw as much as possible at one time. Make good notes on the drawing itself—such as: "blue and gold double", "red star", "brightest on east side", etc.

Under white light back at home, now is the time to make a gorgeous second drawing that you can show proudly at the next astronomy club meeting. Keep the originals of your drawings and make redraws for other uses. Remember, with this technique, you are making a negative drawing. What was bright in the eyepiece is dark on the finished drawing. Practice and this will be second nature.

A trip to a university book store or art supply shop will get you some useful materials. Use a #1 pencil (B in the UK) or charcoal pencil for nebulosity. Smear around with finger or blending stump. I happen to like the blending stump, which is a pencil shape of tightly rolled cardboard that will serve to smooth out the graphite or charcoal on the paper surface. Use a #3 pencil (H in the UK) or a razor-point pen for stars.

Actually, I find that I am using both on the same drawing. The pen works great for bright-field stars and stars embedded in nebulosity, while the hard pencil is best for faint stars. Erase or don't draw in the dark lanes or globules. Get a good eraser, and be careful of smearing the drawing with left-over charcoal or graphite from previous use. I keep a piece of sandpaper and a rag handy to get rid of extra graphite from the blending stump or eraser (See Fig. 6.1).

Making color drawings at home means another trip to the art supply shop. Now your drawing will be on black paper; try not to get paper with too much grain or texture. I like my results with heavy drawing paper. A selection of colored pencils is necessary and I use several blending stumps so as not to smear one color into another. If you use your finger as a smearing tool, wipe it off before each use. Because you are drawing a positive now, black pencil or pen will add dark markings or lanes in the white nebulosity. Before a good color drawing can be done, a very complete field drawing must be available. The colors and where they are located must be documented at the eyepiece before attempting a final color sketch.

I have found that making bright stars in the field of view is difficult. Having tried both making big dots and adding diffraction spikes, it seems that neither technique is satisfying. The good news is that few deep-sky objects lie within the field of view of really bright stars.

Storing your treasured drawings and observations can become a factor as you gather more and more materials over the years. I have found that manila folders, arranged by constellation, are best. After a modest investment to start and few hours filing, now I can easily find a drawing or observation from among the objects I have observed and sketched.

The size of your telescope does not matter. I have some excellent drawings of wide-field objects from my binoculars or finderscope. So, don't listen to that little voice in your head that says you can't do this. Spend a little money, get some of the right materials and you can be drawing heavenly bodies in no time.

Hoist with My Own Petard

Voluminous notes and drawings are not always a blessing, however. While observing in Corvus, I looked up NGC 4027, a rather bright galaxy. When the magnification was raised to 150× there was a 13th magnitude star near the core. I estimated the position angle at 30° in relation to the core and made a sketch.

All of this was done while wondering if I had discovered a supernova. So, I took careful notes, thinking that I would return home to a triumphant celebration of my achievement for being the first to spot this exploding star.

I thought of just e-mailing Brian Skiff at Lowell Observatory with my results. Then I stopped and realized I should check my modest library to see if there is an image of this galaxy for comparison. Nothing in *Burnham's,* nothing in *Visual Astronomy of the Deep Sky;* I had already checked the image in the *Deep Space CCD Atlas* by John C. Vickers.

Then I remembered the *Night Sky Observer's Guide,* pulled volume 2 off the shelf and opened to NGC 4027 in Corvus. There on page 119 was a drawing from many years ago that included the star quite prominently. So, my dreams of glory were shattered in an instant. A quick glance at the artist's name, and I saw—*it was me!*

Chapter 7

How Can I Find All These Deep-Sky Goodies When the Sky Is So Huge?

An entire chapter of this book is dedicated to capturing and recognizing what is in your telescope. That is because being confronted with finding your way around the dome of the sky can take the fun out of astronomy very quickly. Being able to point your telescope accurately at the target you are seeking will define your ability to have some joy in your observing: no find, no fun. Seems obvious.

Watching the Sky

Before we leap into a discussion of star charts, finder-scopes and setting circles, it is important to find your way around the sky with just your eyes. Much of observing the Universe demands that you have rudimentary knowledge of how the sky is constructed and how it moves. At the beginning of this book I said that you should be at a level of accomplishment that allows you to find the brightest objects. I am assuming you have found the Andromeda Galaxy and the Orion Nebula and a few others.

The task for you to take on is to observe the sky. Don't just glance. Spend some time watching how the dome of the sky rotates. Don't get distracted by the telescope, just sit in a chair for a few minutes and view the sky. Come back in a couple of hours and notice how the sky has changed.

Notice the path the Sun and Moon take as the months progress. Learn a few faint constellations. With nothing other than the centerfold chart out of an astronomy magazine and a red flashlight, much can be observed about the Big Picture.

After a few months of observing the entire vault of the heavens and watching the sky dance around your home, you will have educated yourself about a part of Nature that few modern people know. Life in the Big City just does not lend itself

S.R. Coe, *Deep Sky Observing*, The Patrick Moore Practical Astronomy Series,
DOI 10.1007/978-3-319-22530-2_7

to observing the sky. Much will happen to change your life in this period of time. Friends will say, "Are you going out with the telescope this weekend?" You will answer, "No, it's Full Moon." Then they will look at you as if you were a mental patient. Only a small fraction of people alive today really understand and follow the movements of the sky.

Knowing and following the movements of the sky can acquaint you with a part of your surroundings that can become familiar and enjoyable. Seeing the return of a favorite constellation will start to mark the change in the seasons. Watching the crescent Moon thicken and grow full will remind you that a month is actually a "moon-th". This knowledge can also get you in touch with our ancestors. The Pyramids, Stonehenge and the Aztec temples are all aligned with the movement of the Sun, Moon, planets and stars. All those centuries ago, the ancients felt the need to be in touch with the vault of the heavens. You can join in that fraternity.

Star Charts

Just one size and scale of star chart is not enough for most determined observers of the sky. One set needs to be a few charts which cover the entire sky; with them you can find your way around in general. If the notes about a galaxy say "2° north of Eta Pegasi", I need to know which star is Eta. I can find Pegasus (See Fig. 7.1).

For this wide field level of chart, I like the *Bright Star Atlas 2000* by Wil Tirion. It provides an excellent level of detail, so that you can get to the general area you wish to observe. Facing each chart is a fine listing of all the best and brightest objects to observe in that part of the sky. *Norton's Star Atlas* also covers this scale of star atlas very nicely.

Fig. 7.1 Several scales of star charts are needed to find your way around the sky

The next level of star chart shows a wide variety of deep-sky objects and double stars at a limiting magnitude faint enough to make it useful with a modest telescope. *Sky Atlas 2000* plots stars to magnitude 8 and 2500 nebulae, galaxies and clusters. I used a *Sky Atlas* for many years and it will get you hooked on deep sky.

Uranometria 2000, the *Millenium Star Atlas* and the Herald–Bobroff *Astro-Atlas* are at the next level. In fact, *Uranometria 2000* is the "deepest" sky atlas in my observing box.

That is because I am used to it. Any of these three will do an excellent job, providing you with the location of enough deep-sky goodies to keep you observing for a lifetime.

Regardless of the star atlas you have chosen, make certain that you take the time to determine the field size of your finderscope on that chart. I would also recommend finding the size of your widest-field eyepieces in the main instrument. This will give you an idea how far to move the telescope to hop from one object to another as you find your way around the sky. Without that knowledge you will likely do what I did with my first telescope. After setting up my brand-new 8 in. f/6 in the backyard, I searched for the Andromeda Galaxy for over an hour. After determining the field of the finder, I realized that, on my first night out with a telescope, I thought the main scope and the finder had the same field of view.

I understand that many observers are using a computerized device to use as a star chart. I have gotten good at using a printed chart, but the effect is the same. You still need to know the size of the field of a finder scope and how that relates to the sky. So much has changed, as a man with grey hair, it is fascinating to me with what we call a "phone" in today's world.

Finder Scopes

Anyone who has tried to point a telescope at a particular place in the sky has very quickly learned the value of a finderscope. Attempting to look down a round telescope tube and determine exactly where the scope is aimed is futile and time-consuming. There are several types of finders on the market. The most popular is a small refractor with a crosshair eyepiece which allows the user to accurately determine the place in the sky where the telescope is pointed. The smallest of these refractors are 5 × 24 and 6 × 30 systems. That means they are 5 or 6× magnification and have a 24 or 30 mm objective lens. These small finders are only useful for finding bright objects. If the only objects you are looking to find are the Moon, planets, and bright stars, then a small finder will suffice.

Moving up to an 8 × 50 or 11 × 80 finder will start to show brighter deep-sky objects and will provide an observer with fainter stars. These two sizes of finder are very popular with deep-sky observers and I have an 11 × 80 myself. Besides being able to see dimmer stars and objects, there is another advantage. They make great rich field telescopes. The view of the Sword of Orion in the 11 × 80 with a UHC filter is beautiful. The entire region is nebulous, with M42 showing dark markings and a large loop of nebulosity which leads to Iota Orionis. The Rosette, Lagoon and North America nebulae are also very nice in a small telescope that can show some of the area which surrounds the main object.

The last step up is a small Newtonian telescope as a finder for a larger instrument. I have used a 4.25 in. f/4 as the finder on my old 17.5 in. on a few outings when I was looking for something very faint. The problem here is that even a small Newtonian only has a field of about 2°. That can be a problem unless you have a small finder on top of the large finder.

The one thing that must be done before a finder-scope can be used to its full potential is to align it with the main scope. The two telescopes must point at the same spot so that when the crosshairs are on a location in the finder, the main scope points precisely at the same location. The most common way to do this is a set of rings which support the finder and allow the user to align the finder by adjusting the set screws on the rings. Some aluminum tape around the finder scope's tube will provide material into which the set screws can grab, without scarring the finderscope.

Now for one of my best tips on finder scopes. Do the alignment during twilight. Here is yet another reason to arrive at your observing site before sunset. Use a distant object such as a hill, telephone pole or (dare I say it?) street light. Once it is dark, I am anxious to start observing, so I don't do as good a job as I should.

You didn't think I was going to write about finder scopes and not mention "zero-power" finders? Even though they don't magnify the view, the zero-power (or "unity") finder has made navigating the heavens much easier for many people—me too. A small "bull's-eye" pattern is projected on a piece of glass which is at a 45° angle. The pattern is focused for infinity, so it appears projected on the stars. Easily the most common of these is the Telrad. It is an excellent investment for your telescope (See Fig. 7.2).

Fig. 7.2 The Telrad is a very popular zero power finder. It comes with a mounting plate

Fig. 7.3 The business end of my 13 in., with Telrad and 11 × 80 finder with Amici prism

If you purchase a zero-power finder, be careful with the device. Make certain you return it to its case after every trip. Otherwise, you will eventually break the glass. Because the only electrical current draw on the batteries is a small LED to light the pattern, batteries last a long time. I change mine once a year or so, just because I am guilty by then (See Fig. 7.3).

My 13 in. f/5.6 Newtonian is equipped with a Telrad zero-power finder in conjunction with an 11 × 80 finder-scope. The Telrad lets me make certain where I am starting in the sky and the large finder lets me see dimmer stars to find my exact location. This works very well when using the *Uranometria 2000* star charts. Their limiting magnitude is easily seen in the 11 × 80 and would be only glimpsed in an 8 × 50 finder. I feel much more confident finding my way around with this system than with any other combination I have used in the past. The Telrad and the 11 × 80 finder are installed on opposite sides of the focuser. That way neither of them gets too far under or on top of the tube as I rotate the tube for the most comfortable eyepiece position.

The other tip that I can pass on concerns Amici prism diagonals. This right-angle device will correct the field of view in your finder so that the orientation on the star charts is the same as the view in the finder. It makes your finder into a true monocular (half a binocular). The advantage is that the user does not have to try and reorient the star pattern from the star chart. I found it very difficult to flip the field in my mind; the Amici prism alleviates that problem. The disadvantage is that there are some light losses in the prism, so it works better with the larger finders. Both University Optics and Lumicon offer premium finder scopes with Amici prisms installed. I know that they are expensive, but I find mine well worth the price (See Fig. 7.4).

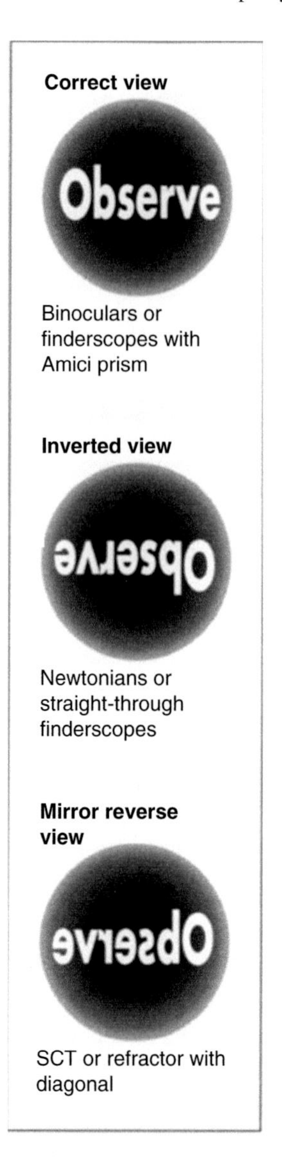

Fig. 7.4 Image orientation

Size of the Field of View

In Chap. 5 I gave you a formula that allows you to calculate the size of the field of view in a particular eyepiece, and it works very nicely. There is also a way to determine the field of view in any optical instrument with a simple drift method. Use a star that is near the celestial equator, which means a star with a small value for its declination. Put that star at the edge of the view and allow it to drift through the field, while you count the number of seconds it takes to cross the field of view. If you wish to be precise, there is always someone around with a new digital watch and they will be happy to show off the stopwatch feature among the 20-odd pushbuttons.

Once you know the drift time in seconds, divide that number by 4 and the result will be the width of the field of view in arc minutes. So, if you timed the drift at 120 s (2 min), now divide by 4 to get the result of a 30 arcmin (0.5°) field of view. Another example: you timed the drift at 40 s, so divide by 4 to get a 10 arcmin field. Any answer larger than 60 arcmin is over 1°.

Knowing the field of view of your eyepieces is imperative if you are going to find your way around on a large, complex star chart. To be confident of where you are on that chart, you must know the field of view of your telescope. I recommend doing the timings for all your eyepieces while in your backyard and then write them in ink on a small index card to have them handy.

Directions in the Eyepiece

While it is no easy skill to master, you *must* be able to determine which directions are which in the field of view. Have some patience and don't give up; this will provide you with the ability to know where you are and where you are going.

Let's do the easy direction first. If you have a telescope with a motor drive system, turn it off. Center a star at 75× to 100× and don't touch the scope for a few moments. Because we live on a planet that rotates on its axis, the star will slowly move toward the edge of the field of view. The star is moving toward the west in that view. Where it is at the instant when it exits the field is the western boundary of the field of view. Opposite that point, where stars are entering the field, is the eastern side of the view. I can remember this because the eyepiece view mimics what happens in the sky—things rise in the east and set in the west. Easy.

I know of two ways to find the north-south line in a telescope. First, find that western point and if you are using a Newtonian or a refractor with no star diagonal, then north is 90° counterclockwise from that point. If you are using a star diagonal with a Cassegrain scope (SCT or Maksutov) or a refractor, then north is 90° clockwise from the western point. If there are even numbers of reflections in your scope, then the movement is counterclockwise from west. An odd number of reflections and north is clockwise from west.

A second method for finding the north–south line is to move the scope slightly and watch the stars in the field. I will discuss this for Northern Hemisphere observers first. Once you have found the east–west line then the north–south line must be perpendicular to it. With your eye to the eyepiece, move the scope gently toward Polaris. The stars will enter the field on the north side of the view. Picture a star map and the field of view working its way north along that map. Doesn't it make sense that more northerly stars are entering the field of view and southerly ones are leaving?

For those of you in the Southern Hemisphere, I know that there is little to help spot the south Celestial Pole. Try moving the scope toward the Small Magellanic Cloud—that is close enough. The stars will be entering the field on the south side. Just remember that the E–W and N–S lines are always perpendicular.

Position Angle

Now that you know the cardinal points in the part of the sky you are viewing, the measure of position angle can provide exact locations and elongations for deep-sky objects. This method is exactly the same as using a magnetic compass on a map. North is always 0° and south is always 180°, east is 90° and west is 270°. If the position of an object is 86° from a star, then it is just a little north of due east. I know that the sailors among my readers already know this; for landlubbers, practice will make this very useful.

Let's do a real example, using galaxies in Virgo. NGC 4388 is 16 arcmin to the southeast from M84. I will assume that you can find M84. Now determine the cardinal points in the field. If you are using an eyepiece that provides a field of 30 arcmin then center M84 and pull the scope a little to the south and then a little to the east and you will be viewing NGC 4388. Good job!

Notice that NGC 4388 is elongated. This is the opportunity to use position angle for its other purpose. That function is to inform the observer of the angle of elongation. Because you know the cardinal points in the view, you quickly realize that NGC 4388 is elongated in an E–W direction. Then look up this galaxy in the *Deep Sky Field Guide* or the Saguaro Astronomy Club database, or log on to the ngcic.org Web site and find that an accurate measurement of its position angle is 92°. Good enough: you are now certain that this is the galaxy you were looking to find.

Putting It All Together

Let's find and describe a nice galaxy as an exercise in how to use position angle and finder systems. To find NGC 2903 in Leo, look up its position in the sky and find 09 h 32.2 m for right ascension and +21° 30′ for declination. That leads to *Uranometria* chart #143 in the first version. Notice that NGC 2903 is almost dead south of a bright star in Leo, Lambda Leonis. Go to the Bright Star Atlas to find where Lambda is located and see that it is in the Head of Leo.

With the star charts open and my red light on, use the zero-power finder to center Lambda in the red circles or on the red dot. Look in the finderscope to make certain that the star is centered. Put a wide-angle eyepiece in the scope, one that gives at least a 30 arcmin field. On *Uranometria* charts each square box is 1° on a side. NGC 2903 is one and a half boxes south of Lambda Leonis.

So move the scope three fields of view to the south and the galaxy should be in view. If not, make certain that all three devices are centered on the star when it is in view. If your eyepiece gives a wider field, then you may need to only move two fields to the south. It seems to me that knowing your field of view is one of the most difficult things to grasp when learning your way around the sky.

When you are looking at a particular chart, try to picture how large the field will be when represented on that chart. Don't expect the entire constellation of Leo to fit in the finderscope. Know that a wide-angle field in the main instrument will be about one square degree on a *Uranometria* or other deep star chart. If you get field size and orientation correct in your mind, you will not get lost nearly as often while trying to find your way among the stars.

Get Polar-Aligned

If you are the owner of an equatorial mount, then take the time to get well polar-aligned. It is not a difficult procedure and once you get good at it, it can be completed quickly in twilight. There are several reasons to get well aligned. First, the scope will track objects only if the polar shaft is aimed correctly. This means that at high-power a small nebula or galaxy will stay in the field of view longer. Also, if you are drawing or making a detailed set of notes the object will be available in the center of the view. Another reason is that if you plan on using setting circles to find your way around the sky, then getting on the pole is important.

We generally travel to about six sites to observe, so there is a history of what those sites look like. Therefore, at each site spot a distant hill or other landmark that is either due north or due south. That way, even if the ball of the Sun is still above the horizon, the scope can be roughly aligned along a north–south line (See Fig. 7.5).

Once the stars are visible and the pole's position can be determined more accurately, then point the axis at the pole. I use a simple polar finder, of which many are available on the market. Alternatively, you can estimate the distance from a pole star to the true pole in your finder once you know the size of the field of view. The polar finderscope will get your main telescope aligned well enough for most nights.

If you are planning to do some piggyback astrophotography or are going to use very high power, then a more accurate alignment is needed. For this accuracy, I use the drift method. The advantage is that you are actually tracking a star with the scope, so you are seeing the accuracy of your alignment, right in the eyepiece. As always the much-maligned southern observer will need to reverse the drift directions. (Hey, you guys—have Eta Carina, 47 Tuc, the Magellanic Clouds and Omega Centauri overhead!)

Put a crosshair eyepiece in your scope, if you have one. If you don't have a crosshair eyepiece, use a power of 150× to 250×. What you will be doing is setting a star in the center of the field and seeing which direction it drifts away from the center as you use the RA drive corrector to keep the star as close to the center as possible. There are two movements of the axis to be checked. (1) Center a star near the celestial equator (a declination that is near zero) and near the meridian (the highest point above the horizon).

Fig. 7.5 A compass will help you align an equatorial mount during twilight, before Polaris is visible. We often ask, jokingly, "Has Polaris risen yet?

If the star drifts northward, the scope axis is west of the pole.

If the star drifts southward, the scope axis is east of the pole.

Now adjust the axis of the telescope a little and then redrift a star to see if the drift is less than before. When the star's drift is the opposite of the previous pass, then you have gone too far and need to move the axis back a little.

(2) Center a star near the celestial equator and the eastern horizon. (I use about 20° up for convenience.)

If the star drifts northward, the scope axis is high (above pole).

If the star drifts southward, the scope axis is low (below pole).

If you have never done this before, then practice in your backyard until you can get a drift that keeps the star centered for at least 5 min, more if you plan on taking photos. It will not be easy the first few times, but practice will allow you to finish this task in twilight and not miss any observing time.

Mechanical Setting Circles

Before I start the conversation about how to use setting circles, please let me emphasize that novices should avoid starting here. This is not an easy skill to master and using setting circles does not teach you how the sky appears and the location of objects to observe. Please learn to star-hop your way to some brighter objects before setting out to use circles.

Virtually all commercially made telescopes today have mechanical setting circles. Some are larger and more accurate than others, but you can get near the object you are looking to find with any of the circles right out of the box. I have never found circles on a portable telescope to be accurate enough to allow me to swing the scope all over the sky and put objects in the field of view. So, I use them to get me close and then star-hop to get on the object I am searching to find.

Here is an example in Orion. I am using *Uranometria* chart #181, but any star chart that includes Betelgeuse and has stars to 7th mag will do. So, turn on the drive motor and align the scope with Betelgeuse. Reading from the edges of the chart, its estimated position is RA 5 h 55 m and the Declination is +7.3°. Move your circles until they read those values, making certain that Betelgeuse is still in the center of the eyepiece.

Now, let's move near to our target, NGC 2169, a small star cluster to the northeast of Betelgeuse. Notice that the star Nu (or 67 Ori) is just north of our quarry. Move the scope so that the circles read the position of Nu, about RA 6 h 7 m and +14.8°. Look in the finder and then the main scope to center this star. There is a nice chain of three fainter stars to the west of Nu; use these to make certain that is the star you have centered. (A reminder: stars to the east have higher right ascension values.) Now we are ready to move to the star cluster NGC 2169. Move the scope a small amount to the east because the cluster is not directly south of Nu. Then move the scope not quite 1° south to the cluster itself. If you are at 100× or so, you should see a pretty bright, but not compressed group of 18–35 stars, depending on aperture and sky transparency. While you are there, see if you can notice why this is called the "37" cluster. Move your head around and at the correct angle the stars form the numerals 3 and 7 (honest!). If you were unable to find this cluster, then start again, align on Betelgeuse and set the circles, hop to Nu and down to the cluster.

Now that we know the scope is well aligned, let's try a more difficult object, NGC 2022. A glance at the position of this little planetary nebula shows that it is east of 40 Ori; the position of the hop star is 5 h 37 m and +9.2. Once on the star, move to the east and get the two fainter stars which point at NGC 2022 in the center of the finder. Look in the main scope and make certain that the stars are in the field. Now move a little more east and a little south to the tiny non-stellar dot—that is it! How much is "a little more"? It all depends on your telescope and eyepiece combination. Once you have found this little planetary, try a higher magnification.

So the technique is to find deep-sky nebulae, galaxies and clusters by each time starting out at an easy-to-find star and then moving the scope to the position of the object. Using the circles, finderscope and low power in the main scope should get you consistently on target.

Electronic Setting Circles

The basic idea of electronic setting circles is having a small computer to help you in pointing the telescope accurately at objects in the sky that you wish to observe. The two movable axes of the telescope each have an electronic detector, called an encoder that provides information about the angle at which the telescope is pointed. The controlling hand paddle allows the operator to enter "Saturn" or "NGC 4631" and the computer calculates in which direction and by how far the telescope needs to move to put that object into the field of view.

Now there are two ways to control the telescope. The computer can inform the operator of the direction and amount of movement and the operator moves the scope until the display reads that the object is in the field. I call this type of control "push to" since the operator pushes the telescope along in the sky and is helped by the computer to line up the scope at the correct location.

The second choice is "GOTO" and in this type of instrument there are motors on the axis of the telescope which will move the scope until it is pointed at the object of choice. One of the advantages of GOTO is pure ego. It is just plain neat that once you pick out something to view and hit the "Enter" button the telescope moves all on its own to line up on the correct place in the sky. If you are at a public viewing session, every 10 year old child will light up with "it finds things all by itself?". Even if that 10 year old child is trapped in a 40 year old body, it is fun to watch.

There is one thing that you will have to get good at once you purchase one of these scopes. You didn't think it was going to be all simple, did you? You must get good at finding a dozen or so of the brightest stars around the sky. When the system starts up it needs the location, time and date. You can give it that from a map and a clock, or some telescopes come with a GPS detector built into the system. Once the computer knows that information it will need to point accurately toward at least two bright stars. So, when the hand paddle reads "slewing to Almach", or another star with that type of name, you will need to know where that is located. Get a low cost star chart or the centerfold of a monthly magazine and they will provide that information. So, you must know the sky somewhat to get these electronic setting circle systems to work.

Many of these computers have a database of things to view included in the hand paddle memory. But, there may be reasons that you are beyond that data. One is when a comet is on your list of things to view. To find a comet I use my planetarium program to give me the Right Ascension and Declination position of the comet for that night and time. Then I enter that position into the hand paddle and call it up once it is dark. Your planetarium program may have to get on the Internet and acquire the orbital elements for that comet so it can calculate the location for that night. There are also websites that will give you the RA/Dec for each day. That is called an ephemeris.

My first GOTO system was a Celestron Nexstar 11, an 11 in. Schmidt-Cassegrain telescope in a fork mount. I fell in love with one at the Riverside Telescope Maker's star party and bought one several months later. I found the optics quite good and electronics perfect. Once I got trained on how to use it, that telescope never missed putting that object into the field of view of my 22 mm Panoptic eyepiece. What a joy.

I also assembled several simple alt-az telescopes that included the Sky Commander set up. Once the encoders were in place and the computer hooked up, it all worked very nicely. I pushed the telescope until the display told me I was at the right location and I was ready to view.

My most recent mount is a Celestron CGEM. It is a modern German Equatorial mount. I have used it for some simple wide field astrophotography and lots of viewing. Again, once I got good at using the system it puts objects in the field all night long. I had also found that I had missed a tracking telescope. It was nice to find that I could look away from the eyepiece for a few minutes and when I looked back the object was still in view. Also, it made drawing what I was seeing much easier. Having the object stay centered in the view really helped when I was sketching what I was seeing.

The Names of Deep-Sky Objects

I am certain that you have very quickly bumped into the wide variety of names that deep-sky objects are given. Let's face it, people have egos and want their name attached to the Universe. So, each catalog lends itself to such verbiage as: "Is this M26?", "I've got NGC 4565 in my scope," "Stock 2 is the big cluster north of the Double Cluster" and "What is the Barnard number for the Horsehead?" It can certainly be confusing.

Two of the most prominent designations are M for Messier and NGC for New General Catalog. They became famous through two very different routes. Charles Messier was a French comet-hunter and one of the world's most prolific in the eighteenth century. He kept finding fuzzy, nebulous objects that were not comets and they confused his comet-hunting efforts, so he started making a list of nebulous, non-stellar objects that are not comets because they don't move. Because his telescopes were small, and rather poorly made by modern standards, he only found the brightest objects in the sky. Therefore, today's observer can be assured that if they have completed the Messier list, then they have seen many of the showpieces of the northern sky (especially if they happen to look a bit like comets in a poor telescope).

The NGC was compiled in an attempt at providing a numerical designation for every object seen to that time (1888). William and John Herschel are the major contributors to this list of about 7840 objects. Beyond the Herschel's 5000 contributions, many observer's had small lists that were all put together in a monumental task by J.L.E. Dreyer. So, the NGC was an effort to cover the entire sky and list all that could be seen in a good telescope. The NGC numbers are assigned by increasing right ascension. With a few exceptions, all the brightest objects got included in the NGC. This is the reason it is still in use today. The IC is the Index Catalog, in which Dreyer added objects found since the NGC was published.

The only types of deep-sky objects that were available, and yet missed by the compilers of the NGC are large star clusters and dark nebulae. Wide-field photography in the beginning of the twentieth century allowed Collinder, Stock and others to discover some wide, bright clusters which have no NGC designation. Many of these are very nice in a rich-field telescope or large binoculars.

Once a modern observer is beyond the catalogs mentioned above, there are many objects to observe, but they are faint. The Herschels and their contemporaries did an excellent job of sweeping the sky for nebulae of all types. So, even though the MCG and UGC contain many tens of thousands of galaxies, most are 14th magnitude and fainter. The same can be said for planetary nebulae. Perek and Kohoutek used the spectrum of these nebulae to discover those in their catalog. However, most are small and/or faint.

Opposite is a list of many of the catalog names given to objects on a large, modern star chart. Some are abbreviated; some use the entire name on the chart.

Abell	George Abell (planetary nebulae)
AGC	Abell galaxy cluster
B	Barnard (dark nebulae)
Basel	(open clusters)
Berk	Berkeley (open clusters)
Bo	Bochum (open clusters)
C	Caldwell (the next 109 interesting objects after Messier)
Ced	Cederblad (bright nebulae)
Cr	Collinder (open clusters)
DoDz	Dolidze–Dzimselejsvili (open clusters)
Dun	Dunlop (globular clusters)
Gum	(bright nebulae)
H	Harvard (open clusters)
IC	1st and 2nd index catalogs to the NGC (all types of objects except dark nebulae)
J	Jonckheere (planetary nebulae)
King	(open clusters)
LDN	Lynds (dark nebulae)
Lynga	(open clusters)
M	Messier (all types of objects except dark nebula)
MCG	Morphological catalog of galaxies
Me	Merrill (planetary nebulae)
Mrk	Markarian (open clusters and galaxies)
Mel	Melotte (open clusters)
M1–M4	Minkowski (planetary nebulae)

(continued)

(continued)

NGC	New general catalog of nebulae and clusters of stars (all types of objects except dark nebulae)
Pal	Palomar (globular clusters)
PK	Perek & Kohoutek (planetary nebulae)
RCW	Rodgers, Campbell and Whiteoak (bright nebulae)
Ru	Ruprecht (open clusters)
Sh	Sharpless (bright nebulae)
Stock	(open clusters)
Tombaugh	(open clusters)
Tr	Trumpler (open clusters)
UGC	Uppsala general catalog (galaxies)
vdB	van den Bergh (open clusters, bright nebulae)
Vy	Vyssotsky (planetary nebulae)
Westr	Westerlund (open clusters)

Chapter 8

Any Other Tips?

This chapter was created so that I can put in a wide variety of tips and techniques that did not fit smoothly into any of the previous chapters. So this part of the book might jump around a bit, but the lessons have been learned the hard way and here's hoping you will not have to do the same.

Let's start with the most important tip of all:

Make a list of things to take with you while observing.

If you leave things at home then all the advice in this book is not very useful. Before creating a list, I had forgotten such things as pencils, paper and a star chart. These are the kind of things I could borrow from a friend at the site. Then one night I forgot one of the eight struts needed to construct the Serrurier truss assembly on my old Dobsonian. It was an entire night of putting up with a very wobbly telescope. This foolish move motivated me to start packing the truck more carefully (See Fig. 8.1).

My list is posted on the wall of the garage and I check it every time before I go observing. Since using this inventory of what is needed to enjoy an evening of observing, I have not forgotten anything important. So list all the things you need in the field and check that list before you leave the house. My checklist is included as a place to start.

Every Time

eyepieces, pencils, note paper, star charts
warm clothes-hood, coat, boots, overalls, gloves
cooler + ice-water, food, soda, paper towels

S.R. Coe, *Deep Sky Observing*, The Patrick Moore Practical Astronomy Series,
DOI 10.1007/978-3-319-22530-2_8

Fig. 8.1 The back of the truck, filled with telescope, ready for an observing weekend. The optical tube is under the blanket for protection from bumps and scratches

binoculars	CB radio
eyeglasses	folding chair
red flashlight	bug spray
radio and tapes	ladder, if public viewing
13 in. Newtonian	
tube assembly + cradle	equatorial mount
counterweight	Telrad + 11 × 80 finder
viewing stand (short ladder) drive corrector	
battery	aperture stop
tool box-tools, C-clamps, weights, nuts and bolts, electrical parts	

Photo System

camera, shutter release cable, stopwatch
guiding eyepiece and batteries

Overnight Trip

sleeping bag, mat and pillow
first aid kit, toiletries
pots and pans, food, stove
for daytime boredom: books, solar filter
sunblock
tarp (plastic sheet large enough to completely cover the telescope)

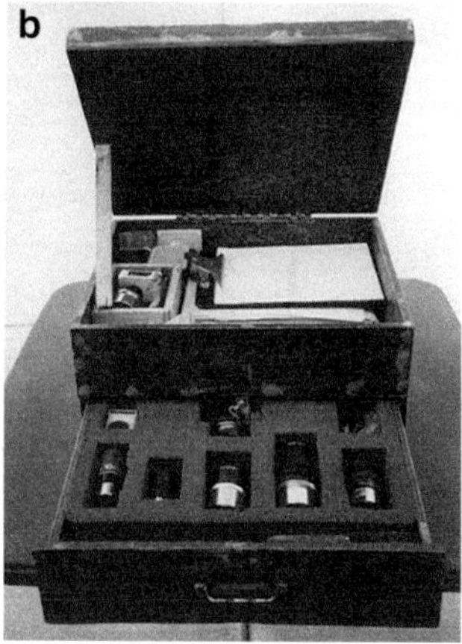

Fig. 8.2 (**a**) An eyepiece and accessories box. The eyepieces are in a drawer along the bottom. (**b**) The top lifts up to store drawing materials and other accessories

I have seen friends arrive at observing sites with no warm clothes, no eyepieces and no drive corrector. On a long expedition to the White Mountains of eastern Arizona, Pierre Schwaar forgot his counterweight. I got to watch this master telescope builder use gaffer tape attaching rocks to the declination shaft and try to adjust them for balance as the night progressed.

Create a box just for telescoping needs (See Fig. 8.2).
Another way to prevent some of the problems of forgetting star charts, eyepieces and pencils is to put them all in one box. An afternoon designing a chart box will save you a lot of grief. Make certain that it is large enough to hold a variety of little things you could easily leave at home.

The photos are of the box my friends and I designed. Note the lip around the edge of the top. This will prevent pencils, eyepieces and flashlights from rolling off the edge of the box while you are working on the top. Also, include lots of room for storage of eyepieces, star charts, extra batteries, clipboards, eyeglasses, pencil sharpeners, paperclips and all the other small but vital paraphernalia that go with being a deep-sky observer.

In recent times the plastics industry has stepped in to fill some of this need. A trip to the DIY store will provide you with lots of choices in plastic boxes from large to small. I use one to carry the reference materials that I use in the field. The box also has room for my finder scopes. I recommend the style which has a clamshell lid with two hinged sections that fold together to form a top. That way there is no separate lid to lose in the dark.

Also, fill a small toolbox so that it contains tools that fit every nut and bolt on the telescope. Obviously, an adjustable wrench will be the most economical way to do this, but make certain it fits all the parts that are difficult to reach. You may need a nut driver or Allen wrench for some special piece of the scope. Carry replacement nuts and bolts, along with any hardware which could fall off the telescope. It is very

Fig. 8.3 The 13 in. from the star's point of view. The battery is hooked up to run the telescope

tough to find telescope parts hiding in the dirt on a moonless night. Some white paint on the tools can make them easier to find. You do not need to buy the most expensive tools for the scope; they probably won't get used often. Wait until the hardware store has a sale and get what you need.

Another very handy item I keep with the toolbox is a cover for the telescope. A large plastic garbage bag works fine. Keep some twine or elastic cords handy to secure it in place. It can be used to cover the scope if the weather turns on you, or if you are going to be at one location for several days.

Keep a battery for the telescope (See Fig. 8.3).
Like many modern driven telescopes, mine runs from a 12V d.c. source and can be powered by an automobile battery. However, being far from a garage at an observing site, I don't like to drain current from a battery that will be needed to start my truck after the observing session is over. To overcome this problem, carry along a second car battery specifically to power the telescope. This provides reliable power for the scope and a backup battery to start the truck if the need arises.

Join a local astronomy club.
I know that some people don't see the benefit in clubs, but please consider some of the advantages of being an astronomy club member. It allows you a chance to chat and exchange information with a variety of other observers. You can hear about the

new supernova or tryout a friend's eyepiece before considering a purchase. There is the fun and camaraderie of enjoying the sky with people like yourself, who love to look at the Universe in all its diversity.

Getting a chance to observe with a wide variety of other telescopes can provide you with a lot of good information. It might allow you to see what a new model can do, or if a new accessory is worth the money to you. It also allows you to see how much detail can be seen in many differing telescopes.

This led to a theory in the Saguaro Astronomy Club, called the "double jump" theory. (Sorry, but this is going to be in inches because there is an intrinsic beauty to the theory that way.) Consider the standard telescope mirror sizes from a catalog: 4, 6, 8, 10, 12, 14, 16, 18 and 20 in. What many years of observing has demonstrated is that the difference in the detail an observer can see between two nearby telescope sizes is minimal. If you are observing a bright star cluster with a 6 in. scope at 100× and next to it is an 8 in. scope at 100× you will see very little difference in the number of stars resolved by these two telescopes.

Next, move on to a 10 in. scope on the same object, also at 100×. Now the difference between the 6 in. scope and the 10 in. one is much more dramatic. Many more stars are resolved in the larger instrument and virtually anyone stepping up to the two eyepieces will see more detail in the larger scope. This is the essence of the "double jump" theory of telescope purchasing. Moving in small size increments will not provide a very noticeable difference in the view. If you are going to consider purchasing a larger scope, then make that "double jump" in size. As an example, if you have an 8 in. telescope, then a 12 in. will be worth the upgrade.

Wear sunglasses until sunset.
The Sun shines bright in Arizona and it takes quite a while for your eye to become dark-adapted. You can help by wearing sunglasses until the disk of the sun is over the western horizon. That way your eyes can get started toward becoming dark-adapted much faster.

Tell a friend where you are going.
Tell your spouse or a friend exactly where you are going to observe and when you plan to return. This is just good sense, especially if you are planning to observe by yourself. Leave a note with a map to the site or a good set of instructions on how to get there. You will be happy when someone shows up because you are overdue.

The Trip to the Multiple Mirror Telescope

A story combining many of the factors discussed so far is the one about the trip my friends and I took to southern Arizona years ago. Our plans included observing from the home of the MMT, Mount Hopkins. We were not going to go unless the weather promised to be good for observing, because it is a 4 h drive. We started checking the weather reports during the middle of the week and by Thursday night were convinced that it would be clear. I called one of my friends to inform him of our plans.

Bill Anderson, A.J. Crayon and I started south on Friday morning. We used the CB radios to chat about what we planned to observe and keep each other informed along the way. Our small convoy made its way to Mount Hopkins without mishap, arriving with several hours of daylight left to search for a site.

We found that the site which the staff of the MMT has erected with picnic tables and telescope pads was not useful to us. There is a large ridge line to the south which blocks out Omega Centauri and southern Scorpius, where we planned to observe. A half-hour or so from this spot was a much better location at an altitude of about 2800 m (8500 ft). We parked there and set up the telescopes in preparation for a great evening. Arriving early had provided us the time to look for a site better suited to our needs.

It turned out to be one of the best nights any of us have ever experienced. A.J. and I have used the 10 point system of rating the sky's seeing and transparency for many years. This was one of only a dozen nights we have rated at 10. I completed all the objects on my observing list for the night and enjoyed several periods of just sitting and enjoying the beauty of the Milky Way. After a spectacular viewing session, we collapsed into our sleeping bags.

After dreaming about having many nights like this one, we awoke at noon to find the weather had changed dramatically. A large portion of the southern sky was a very dark cloud bank, coming straight at our location. The three of us were deciding to start tearing down the telescopes. Suddenly, it started to hail! Not very large hail stones, but large enough to convince three astronomers that this was not a very good place to be. To this day, I think the Muse of Astronomy was telling us that we had our extraordinary night, now leave.

Enjoying Observing Trips in a Motor Home

As I mentioned, I have been traveling in my 30 foot Class A motor home for the past 6 years. It has been a joy and I can highly recommend it to anyone who wishes to make it easier to have fun under the stars. Having so many of the amenities available allows you to relax, get some rest and be ready when the Sun goes down and the stars come out. My setup includes a real bed, a microwave oven, a refrigerator and a toilet. There are comfortable chairs and a couch that will allow you get out of the Sun and wind during the day. Also rain, if that happens.

I have had the fun of traveling to a variety of astronomy events in the western United States and that means meeting a variety of old friends and making some new ones. You can enjoy much of what I have seen with a towed trailer (caravan in Europe).

Fig. 8.4 Motor home and the stars

Chapter 9

What Can I Observe in Galaxies Beyond the Milky Way?

These huge systems of millions of stars, along with gigantic clouds of gas and dust, have much detail to show a careful observer. The first half of the twentieth century saw tremendous advances in the tools of the professional astronomers; telescopes, photographic film and spectroscopic detectors. This allowed our knowledge of the size and composition of the Universe to grow in leaps and bounds.

Walter Baade used these new techniques to determine that there are two very different ages of stars, which he labeled Population I and Population II. The Population I stars are young and located in spiral arms of galaxies. When you look out into the night sky and see the constellations formed by the brighter stars near to Our Sun, those are mostly Population I stars. Population II stars are older stars located all through a galaxy, but gather in large numbers in the core and in globular clusters.

Classifying Galaxies

Edwin Hubble created a scheme of classification for galaxies that has retained its essential features to this day. By photographing many different galaxies with the 100 in. (2.54 m) telescope on Mount Wilson, he created the famous "tuning fork "diagram, which classifies galaxies according to their shape.

Elliptical galaxies form the handle of the tuning fork, and the *spiral galaxies* are the two branches. The S0 type is at the branch of the tuning fork. At first, it was thought that this might be an evolutionary diagram, so that galaxies changed their

S.R. Coe, *Deep Sky Observing*, The Patrick Moore Practical Astronomy Series,
DOI 10.1007/978-3-319-22530-2_9

shape over vast stretches of time. However, modern results have dashed that theory and now it is known that the amount and composition of material and rotational energy at the formation of a galaxy will determine its eventual size and shape.

On a quite smaller scale the formation of a galaxy and the tossing of a pizza crust demonstrate the same laws of physics. Start with a ball of material, spin it round and round, and it becomes a flattened disk with some lumps in it.

Elliptical galaxies present a smooth ellipsoid shape and are composed of Population II stars. They are rated from E0 (perfectly circular) to E7 (highly elongated). In general they present a smooth progression in brightness from a bright core outward. Galaxies with these elliptical shapes are the most common type within the visible Universe. They often comprise the bulk of the galaxies within groups or galaxy clusters. Elliptical galaxies have a wide range in total brightness. The dwarf ellipticals are very faint and exhibit a low surface brightness. However, the giant ellipticals at the center of galaxy clusters are the largest and most massive galaxies.

Spiral galaxies have an obvious core and arms stretching out from the middle. You live inside a spiral galaxy, the Milky Way. That delicate band of light across the night sky is our "star city" within the cosmos. There are three main divisions of spiral galaxies, given by a lower-case letter a, b or c. Spiral galaxies of type Sa have a very large core and small arms that are wound close to the center. Type Sb spirals have an intermediate-size core and quite prominent spiral arms that are farther from the central region. In type Sc spirals the core is quite small and the arms dominate the area presented by the galaxy, unwound far from the nucleus. So, two features of a spiral galaxy determine its classification: the relative size of the core and how tightly wound are the spiral arms.

Barred spiral galaxies have a distinctive bright central feature noticeably longer than it is wide, hence the term "barred". The spiral arms begin at the ends of the bar and loop out from there. There is often a bright core within the bar. These galaxies also use the lowercase letters to designate the relation between core size and distance of the arms from the core. For example, a SBc galaxy would be a spiral with a bar (SB) that has a small core and arms that are massive and far away from the center (c).

Irregular galaxies (I) are an extension of the spiral galaxy class, but they are poorly organized and have little symmetry. There are no regular features in their form or brightness distribution.

A fourth galaxy type is *lenticular,* designated (S0); this is one which shows characteristics of both spirals and elliptical galaxies. They are basically an elliptical in which a central bulge, much like a spiral, can be detected.

Several refinements have been added to this general scheme; intermediate types are given a two-letter designation, such as Sab for a spiral whose pattern falls between Sa and Sb. A spiral with very wide arms and a tiny core is an Sd. Spirals with an S-pattern to their arms have a lower-case s added in parentheses, while ones with a ring pattern get an r. The Magellanic Clouds have short, barely apparent spiral arms, and galaxies like them are given a lower-case m.

Hydrogen II Regions

We will spend time with the gaseous nebulae on a later chapter, but you should realize that the arms of spiral galaxies often exhibit a spotty or mottled texture. This means that the observer is seeing large clouds of gas and dust that are being ionized by hot stars. These areas are called H II (pronounced "H two") regions.

Seeing Galaxy Details

Now let's put it all together. What can be seen in external galaxies? First, the general shape and brightness. Elliptical galaxies will show a smooth brightness distribution from brightest core to outer faint regions. Spirals have a bright central bulge and much fainter arms—just like Our Milky Way. Isn't the core in Sagittarius the brightest part of that band of light? Irregular galaxies show no real organization.

Any time you see a smooth portion of a galaxy, you are looking at Population II stars; the spotty, uneven arms are Population I. The hot and bright stars and gaseous H II regions can be seen as mottling in the arms of spirals. The Milky Way also exhibits these characteristics; look up and down Our Galaxy from a dark site—don't you see clumps of bright areas within a smooth background? That effect is mottling.

The rest of the chapter is a selection of galaxies which will exhibit a variety of sizes, brightnesses and details. S means the rating for the seeing and T is the rating for the transparency or contrast for that night. These ratings use the system detailed in a previous chapter. My notes are included for each object, along with some drawings and photos. All the drawings are mine. I can say that I really enjoy drawing what I see. It makes me pay more attention to what detail is available at the eyepiece. As I said in the chapter on drawing, give yourself some time and patience to get good at sketching, it is worth it.

I really appreciate the efforts of several astrophotographers who helped by providing shots of many of these objects. Their names are next to each photo. The wide field images with no name associated are by me.

This note is particular to the second edition. I understand that some of the photographs from the first edition are mediocre and I did replace some of those with new images made with modern equipment. A few of the images were made by friends of mine who have passed on, Pierre Schwaar and David Healy in particular. I left their photos in this edition as a tribute to them. If you need up to date shots of these objects they can be found in books or on the Internet. I hope you understand.

Object	**NGC 247**
Other names	**HV 20**
Type	**GALAXY**
Mag	**9.1**
Size	**20′×7′**
Class	**SAB(s)d**
Surface brightness	**14.4**
Constellation	**Cet**
RA	**00 47.0**
Dec	**–20 45**
Tirion	**18**
U2000	**306**
Description	**F, eL, vmE 172°**

17.5 in. f/4.5 S=7 T=7 Mediocre site
Dec. 1981 100×—Big blob, no detail.

17.5 in. f/4.5 No rating for the night Good site
Sept. 1983 100×—large, pretty faint, mottled over entire face.

18 in. f/6 No rating for the night Good site
Nov. 1986 No magnification recorded—somewhat bright, large, much elongated, star at one end, not much brighter in the middle.

13 in. f/5.6 S=5 T=6 Good site
Nov. 1988 100×—Pretty faint, very large, very much elongated 4×1 and very little brighter in the middle, somewhat mottled in moments of good seeing.
Trying 150× does not bring out any more detail on this rather mushy night.

My notes from this object really demonstrate that taking good notes takes practice. The early notes are not much help for either demonstrating what the object looked like in the eyepiece or providing enough information so that you could return to the object with certainty. Don't just glance at an object—spend some time with it and make certain that you have seen all there is to see.

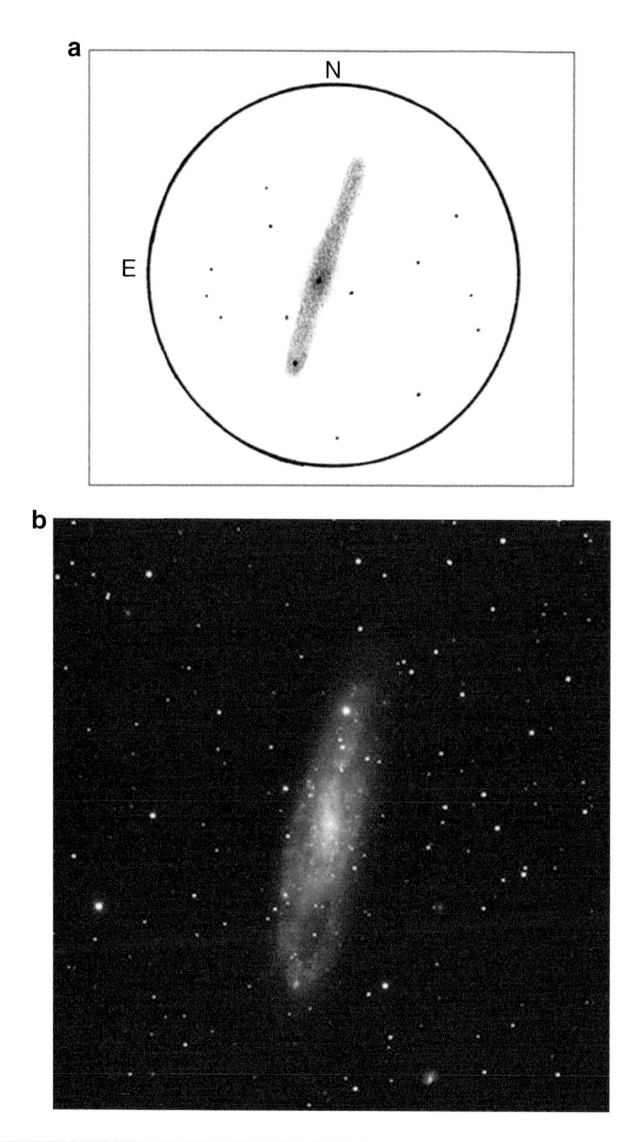

Fig. 9.1 (**a**) 13″ f/5.6; NGC 247; FOV 30′; MAG 100×. (**b**) 8″ Celestron SCT at F/6 *Photo: Scott Rosen*

So for the rest of the objects in the book, I will not subject you to this tedious demonstration of the fact that my early notes generally say "Wow!" or "yuck".

Object:	**NGC 253**
Other names:	**HV 1**
Type:	**GALAXY**
Mag:	**7.2**
Size:	**25′×7′**
Class:	**Scp**
Surface brightness:	**12.7**
Constellation	**Scl**
RA:	**00 47.5**
Dec:	**−20 18**
Tirion:	**18**
U2000:	**306**
Description:	**!!vvB, vvL, vmE 54°, gbM**

11×80 binoculars S=7 T=7 Excellent site

11×—Bright, large, very elongated, much brighter middle. This galaxy is easy to see in the big binoculars from a dark site. There are two stars near to the south side of NGC 253. The core is seen as elongated, even at this low power.

6 in. f/6 S=7 T=8 Excellent site

25×—Very bright, very large, very much elongated 4×1, much brighter in the middle.

40×—Great view; this lovely galaxy takes up the central third of the field of view. There are several dark markings seen across the face of this galaxy. The bright middle is also elongated.

13 in. f/5.6 S=9 T=9 Superior site

Even in 10×50 binoculars NGC 253 is bright, large, much elongated and somewhat brighter in the middle. There is one star on the northeast end and three fainter stars on the southwest end that frame the galaxy.

60×—The 38 mm giant Erfle eyepiece shows this galaxy as very bright, very large, very much elongated

4×1 in PA 45, much brighter middle, very mottled, beautiful convex lens shape, many dark markings in arms.

100×—Excellent view, stellar nucleus evident about 20 % of the time; seven stars are involved with in the galaxy; two pretty bright oval patches in southwest arm. There are many dark lanes with swirls and rifts prominent throughout the galaxy, the most mottling I have ever seen.

150×—Core is gradually sculptured at its edges by dark lanes; an oval bright core has at the center a tiny stellar nucleus.

220×—The core has a fascinating interplay of light and dark, but no spiral structure, more like a surrealistic painting.

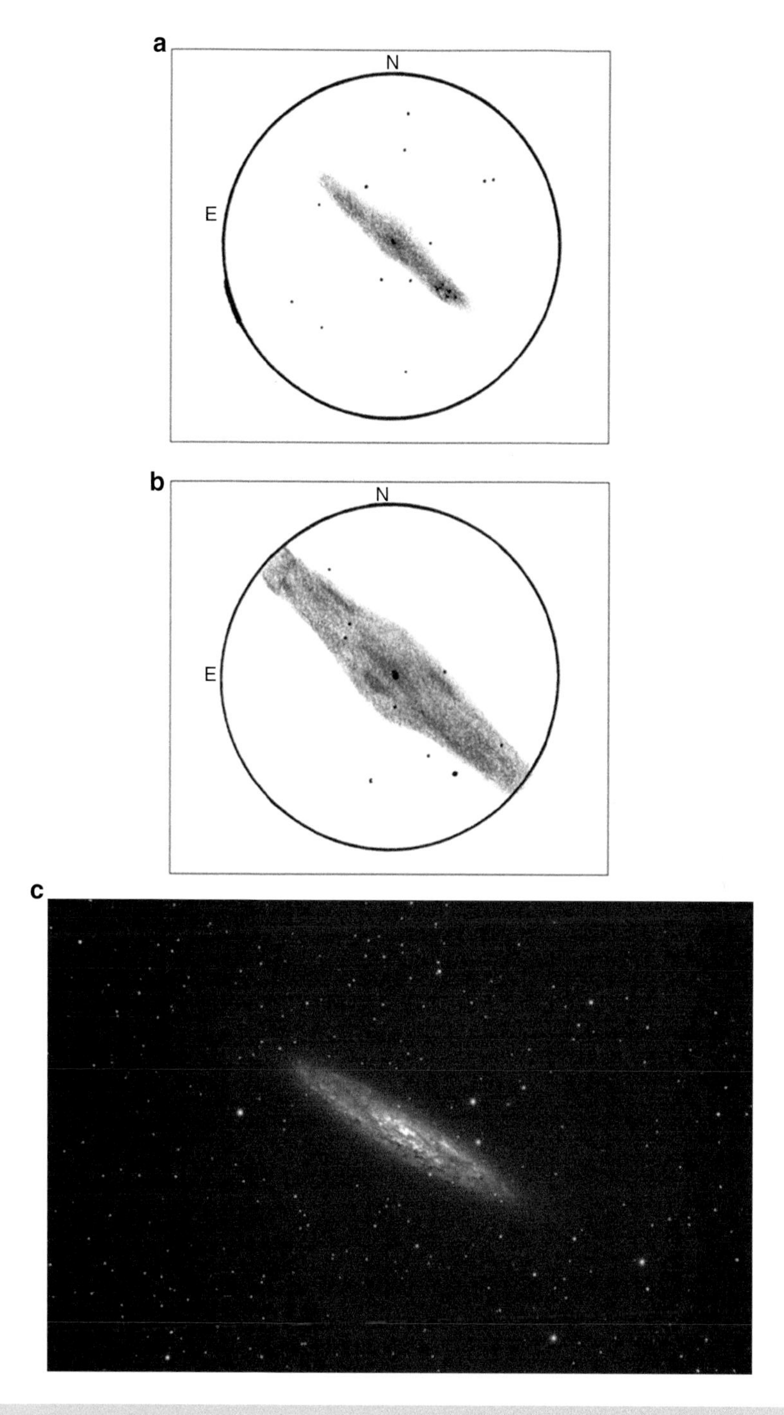

Fig. 9.2 (**a**) 13″ f/5.6; NGC 253; FOV 1.1°; MAG 60×. (**b**) 13″ f/5.6; NGC 253; FOV 20′; MAG 150×. (**c**) TEC 140 f-7 QHY8 camera. *Photo: George Kolb*

In galaxies where much detail is evident, try a variety of magnifications so that you can see all there is to see. Also noteworthy is the fact that this bright galaxy was discovered by Caroline Herschel, sister of William, during one of her comet searches from southern England. She found several deep–sky objects as well as comets and was one of the first women of science.

Object	**IC 1613**
Type	**GALAXY**
Mag	**9.2**
Size	**11′×9′**
Class	**I**
Surface brightness	**15.0**
Constellation	**Cet**
RA	**01 04.8**
Dec	**+02 07**
Tirion	**10**
U2000	**217**
Description	**F, eL**

This galaxy is a lesson in what is meant by "low surface brightness". Notice that even though the total magnitude of this object should make it easy for the 6 in. telescope on a very good night, it was invisible. Also, it was not easy in the 13 in. on an excellent night. So, even though the combined brightness of this irregular galaxy is 9.2, when spread out over the entire galaxy the surface brightness is only 15.0. That is significant because it means that this object provides very little contrast and will not stand out clearly, even if the sky background is dark. I chose to include it in this book for that reason and one other. It is one of the few irregular galaxies that can be seen in amateur telescopes. Every observing list needs a few challenging objects and this is certainly one of those.

6 in. f/6 S=7 T=8 Excellent site

40× and 65×—Nothing seen at the position of this galaxy.

13 in. f/5.6 S=7 T=9 Excellent site

100×—Very faint, large, a little elongated 1.2×1 in PA 90, very, very little brighter in the middle. There is a 12th mag star on the west side, with four very faint stars involved across the face. The entire galaxy is very grainy. Turning on the dim red flashlight to draw this galaxy makes it disappear. The only way to draw it is to memorize the field stars and the position of the faint glow and then put them on paper. It takes several minutes after exposure to the red light to recover enough night vision to detect the galaxy.

150×—Higher power does not help this low-surface-brightness object.
This magnification actually lowers the contrast and makes it more difficult to see.

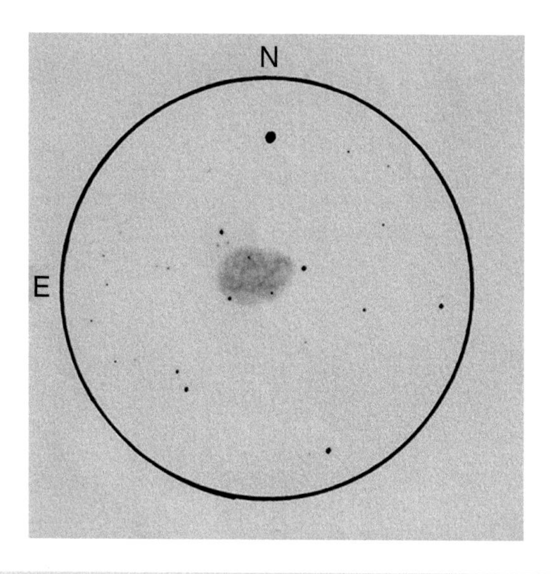

Fig. 9.3 13″ f/5.6; IC 1613; FOV 30′; MAG 100×

Object	**M33**
Other names	**NGC 598**
Type	**GALAXY**
Mag	**5.7**
Size	**73′ × 45′**
Class	**Sc**
Surface brightness	**14.2**
Constellation	**Tri**
RA	**01 33.9**
Dec	**+30 40**
Tirion	**4**
U2000	**91**
Description	**eB, eL, R, vgbMN**

6 in. f/6 S=7 T=8 Excellent site

25×—Bright, very large, elongated 2×1 N-S, bright middle.

40×—A nice view of this galaxy; it takes up about half of the field. M33 gets larger with averted vision. NGC 604 is an emission nebula to the north of the core. Adding the UHC filter gives the nebula some size, a tiny round glow.

13 in. f/5.6 S=9 T=10 Superior site

From the darkest sites, M33 is visible to the naked eye; it is one of my tests for an excellent evening. It appears about 1/3 the size of the Andromeda Galaxy. In the 11×80 finder it is bright, pretty large, has a brighter middle and is afloat in a rich field of stars.

Moving up to the 13 in. at:

100×—Very bright, extremely large, elongated 2×1 in PA 0, much brighter in the middle.

165×—Several H II regions are visible. Most prominent is NGC 604, an elongated glow located 10′ from the nucleus near the tip of the northeast arm. There is a beautiful ‘S’ pattern superimposed over a background glow with a bright core in the center. The entire face of the galaxy shimmers with mottling.

220×—brightest globular cluster in this galaxy can be seen as a stellar point of about 14th mag.

36 in. f/5 S=6 T=8 McDonald observatory Excellent site

180×—With the big telescope, the detail that is seen within the arms of this face-on galaxy is equal to what can be photographed with the 48 in. Schmidt camera. I would not even try to draw it. The curved arms are filled with bright spots from the core to the tips of the spiral. I counted 20 bright areas in the southern arm. I can see the difference between Population I and Population II areas in the galaxy! The stars in the core form a smooth surface to the central section that is light yellow, whereas the arms are splotchy and bluish. The central bright region has an almost stellar core.

Sometimes aperture just wins out. I dearly love my 13 in. scope and have seen much with it. However, when the size of the mirror is almost tripled, a lot more detail can be seen. There is an old saying, “never observe with more aperture than you can afford”.

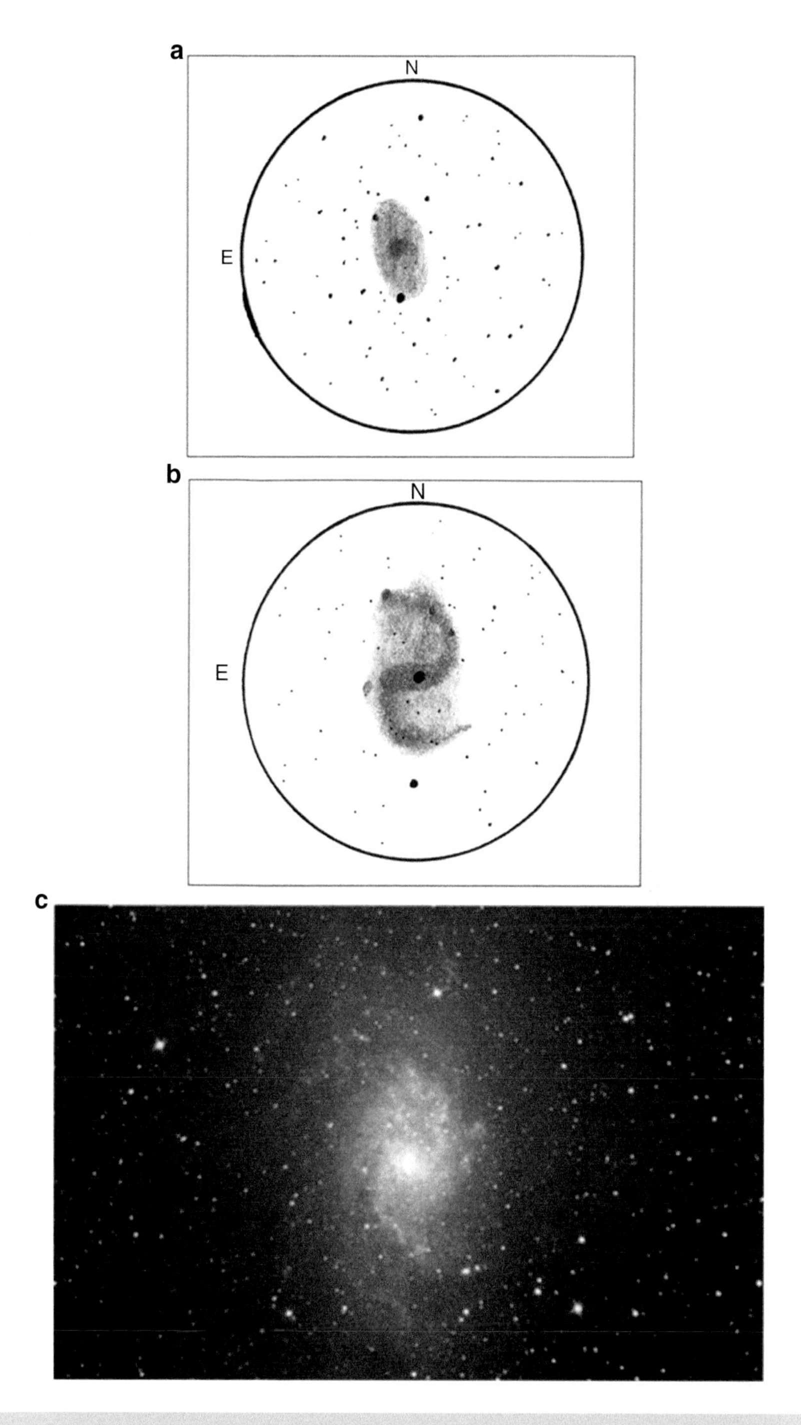

Fig. 9.4 (**a**) 6″ f/6; M33; FOV 1°; MAG 40×. (**b**) 12″ f/5; M33; FOV 25′; MAG 150×. (**c**) 12″ f/5; 1 h exposure *Photo: Chris Schur*

Object	**NGC 772**
Other names	**H I 112**
Type	**GALAXY**
Mag	**10.3**
Size	**8.0′ × 5.0′**
Class	**Sb**
Surface brightness	**13.9**
Constellation	**Ari**
RA	**01 59.4**
Dec	**+19 00**
Tirion	**10**
U2000	**129**
Description	**B, cL, R, gbM, r**

6 in.	f/8	S=6 T=7	Good site

90×—Pretty faint, pretty large, elongated 1.5×1, little brighter middle. Averted vision makes it larger but still pretty low surface brightness.

13 in.	f/5	S=6 T=7	Good site

100×—Pretty bright, pretty large, gradually brighter in the middle, stellar nucleus just seen, elongated 1.5×1 in PA 135. The arms of this face-on galaxy are seen as a glow around the core and this glow is very mottled. Companion galaxy (NGC 770) is pretty faint and small.

165×—Higher power does not show off any more detail in NGC 772; actually the arms disappear.

Sometimes higher powers don't provide any more detail, but you need to try. Be willing to experiment. I certainly have favorite eyepieces, but don't observe with only one magnification all night long (See Fig. 9.5).

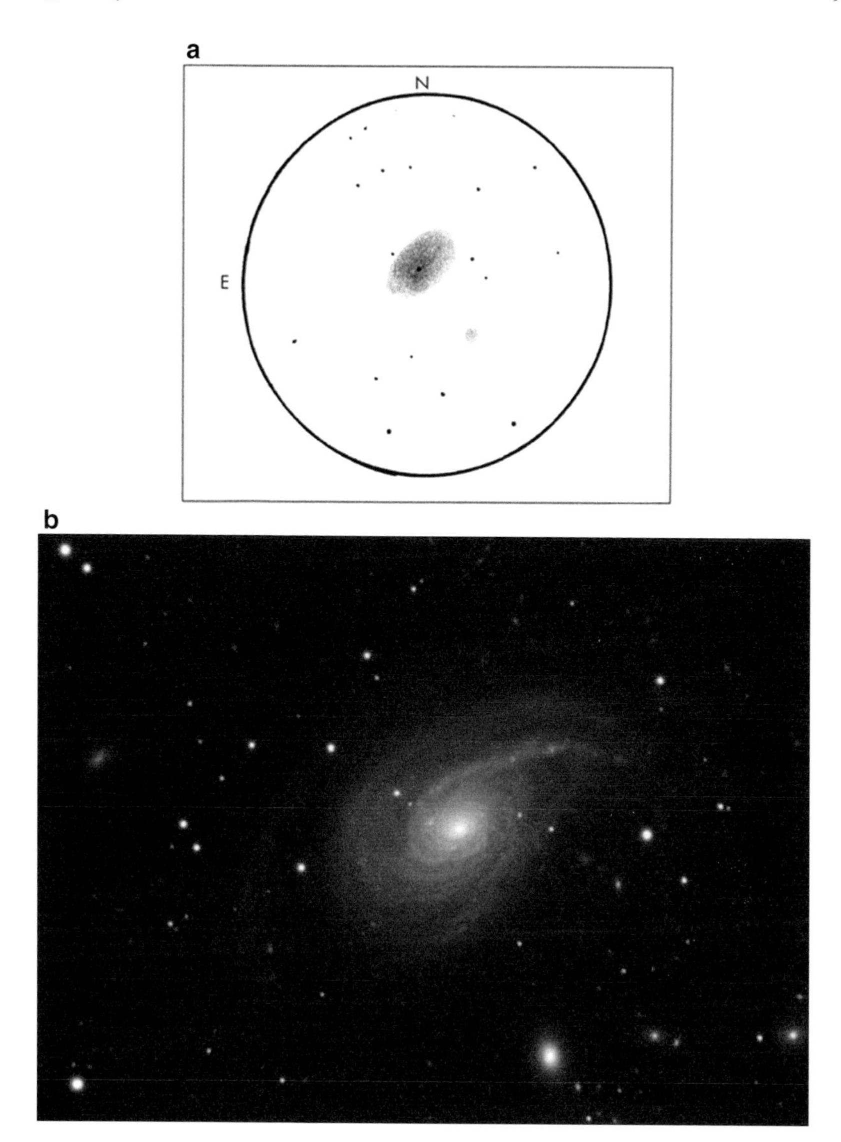

Fig. 9.5 (**a**) 13″ f/5.6; NGC 772; FOV 25; MAG 150×. (**b**) C14 at f7,6 Atik 314L+ camera. *Photo: Parijat Singh*

Object	**NGC 891**
Other names	**HV 19**
Type	**GALAXY**
Mag	**9.9**
Size	**14.0′ × 3.0′**
Class	**Sb**
Surface brightness	**13.6**
Constellation	**And**
RA	**02 22.6**
Dec	**+42 21**
Tirion	**4**
U2000	**62**
Description	**B, vL, R, vmE22**

4 in. f/8 S=8 T=8 Excellent site

60×—Faint, large, very little brighter middle, much elongated 3×1, low surface brightness, averted vision helps, it makes this galaxy more prominent.

6 in. f/8 S=7 T=8 Excellent site

90×—Pretty bright, large, much elongated in PA 30°, much brighter in the middle. The famous elongated galaxy has a dark lane that is off center with the SW side much brighter than the NE. Averted vision shows a dark lane much better and there is a double star near the core that has one component on either side of the dark lane.

17.5 in. f/4.5 S=8 T=8 Excellent site

100×—Pretty bright, pretty large, very elongated 5×1, central bulge obvious.

150×—A thin dark lane almost splits this edge-on galaxy down the middle and the outer arms show sculptured detail, which is more prominent at 200×. Averted vision brings out the dark lane more prominently.

Even though this famous galaxy has been used as an example of an edge-on galaxy in many books, you must avoid the trap of knowing too much before you observe. In a long-exposure photograph it would seem to be very bright and obvious. However, it is a fairly low-surface-brightness object and you might need averted vision to see the dark lane. So, don't expect a blazing galaxy (See Fig. 9.6).

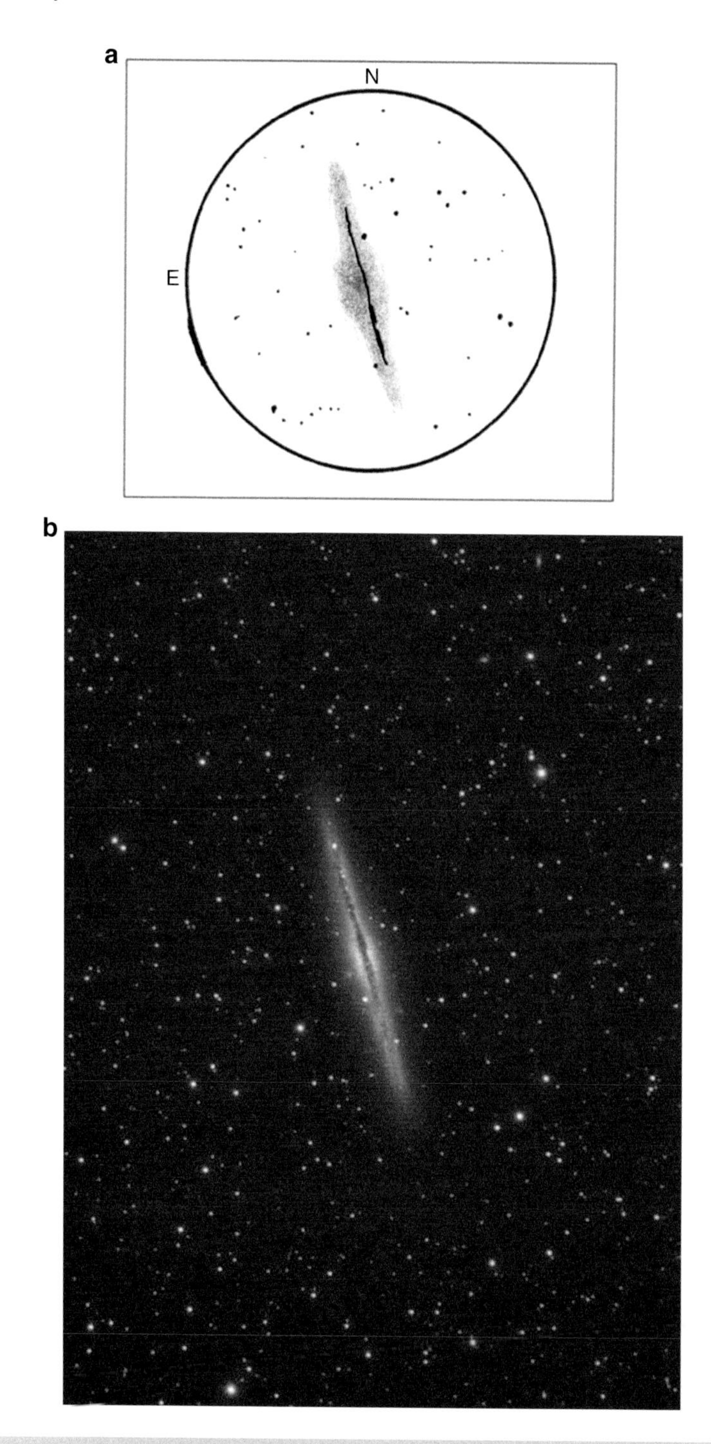

Fig. 9.6 (**a**) 13″ f/5.6; NGC 891; FOV 251; MAG 150×. (**b**) TEC 140 f-7 QHY8 camera. *Photo: George Kolb*

Object	**NGC 1232**
Other names	**H II 258**
Type	**GALAXY**
Mag	**9.9**
Size	**8′×7′**
Class	**Sc**
Surface brightness	**13.9**
Constellation	**Eri**
RA	**03 09.7**
Dec	**–20 34**
Tirion	**18**
U2000	**311**
Description	**pB, cL, R, gbM, r**

4 in f/8 S=6 T=8 Good site

60×—Faint, pretty large, round, gradually brighter middle and very little elongated overall. Using averted vision makes it larger.

6 in. f/8 S=6 T=8 Good site

90×—Pretty bright, pretty large, much brighter middle, round. Averted vision doubles the size, the outer arms are much fainter than near the core. There is a hint of mottling in the arms.

13 in. f/5.6 S=8 T=9 Excellent site

100×—Bright, large, round and gradually much brighter in the middle.

165×—Higher magnifications makes the arms appear very mottled and two H II regions in the arms can be spotted, one to the north and one to the southeast of the core. The core itself is elongated 1.5×1 in PA 90. A tiny companion galaxy (NGC 1232A) can be seen to the east.

220×—Too much power, arms start to disappear.

13 in. f/5.6 S=5 T=6 Mediocre site

100×—Pretty faint, pretty large, round and much brighter in the middle. Low surface brightness makes this object disappear at 135×

13 in. f/5.6 S=8 T=9 Excellent site

100×—Bright, large and gradually much brighter in the middle.

165×—Higher magnifications makes the arms appear very mottled and two H II regions in the arms can be spotted, one to the north and one to the southeast of the core. The core itself is elongated 1.5×1 in PA 90. A tiny companion galaxy (NGC 1232A) can be seen to the east.

220×—Too much power, arms start to disappear.

Use those precious excellent nights to good advantage. Either look for never-seen-by-you objects or go after details within a galaxy that has a previous observation during a mediocre observing session (See Fig. 9.7).

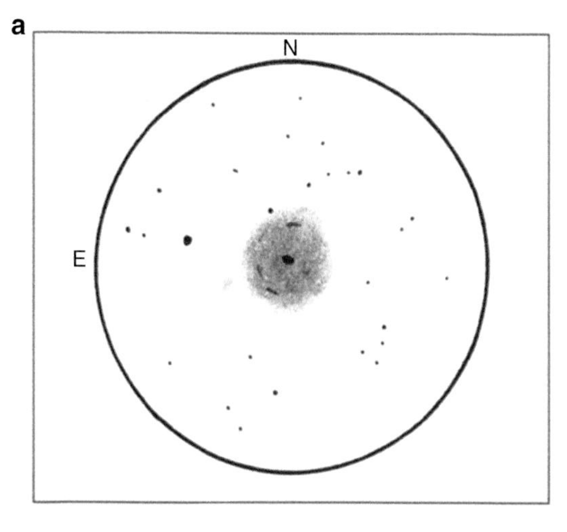

Fig. 9.7 (**a**) 13″ f/5.6; NGC 1232; FOV 25′; MAG 150×. (**b**) C-14 at f/6; 1½ h exposure. *Photo: David Healy*

Object	**NGC 1365**
Other names	**ESO 358-17**
Type	**GALAXY**
Mag	**9.5**
Size	**11′×6.2′**
Class	**SBb**
Surface brightness	**14.1**
Constellation	**Eri**
RA	**03 33.6**
Dec	**−36 08**
Tirion	**18**
U2000	**355**
Description	**!!,vB,vL,mE,BN**

This is one of the best examples in the sky of a barred spiral galaxy. Recently, scientists have discovered that our Milky Way is also a barred spiral. We are just learning why these central bars form, it seems to have something to do with the black hole in the core of the galaxy is spewing out material. They certainly have a bizarre and memorable shape.

6 in. f/8 S=7 T=8 Excellent site
60×—Bright, large, elongated 2×1 with much brighter middle. Averted vision hints at barred spiral structure, but I know it is there so my brain may be "filling in" the view.

13 in. f/5.6 S=7 T=8 Excellent site
150×—Bright, pretty large, much brighter middle, bar structure is pretty easy at this power. The bright middle section of this galaxy is elongated along the same PA as the bar. At higher power the detail is not as easily seen.

25 in. f/5 S=7 T=8 Excellent site
330×—Very bright, pretty large, spiral structure obvious. The central core is cut by a dark lane, it is seen about 50 % of the time. The core is elongated 1.8×1 and the dark lane cuts the core in 1/3 and 2/3 pieces. I can rather easily see several bright knots and some mottling in the arms which extend out from the central bar. This observation is with Dave Kreige's 25″ from McDonald Observatory in Texas.

(See Fig. 9.8)

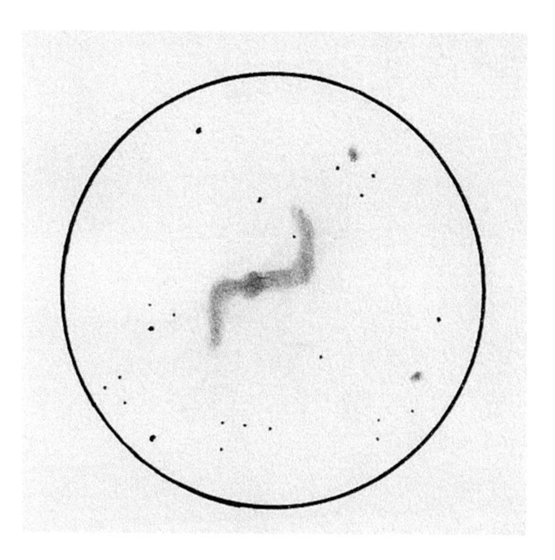

Fig. 9.8 25″ f/5; NGC 1365; FOV 12′; MAG 330×

Object	**NGC 2403**
Other names	**UGC 3918**
Type	**GALAXY**
Mag	**8.5**
Size	**24.3′ × 11.8′**
Class	**SBc**
Surface brightness	**14.4**
Constellation	**Sex**
RA	**07 36.8**
Dec	**+65 36**
Tirion	**1**
U2000	**21**
Description	**!! cB,eL,vmE,vgmbMN**

6 in. f/8 S = 6 T = 7 Good site

90×—Pretty bright, large, much elongated 2.5 × 1 in a PA of 135°, with a gradually much brighter middle. There are three stars involved, one near the core. There is a hint of some light and dark areas in the arms.

13 in.	f/5.6	S=7 T=9	Excellent site

135×—Bright, large and elongated 1.5×1 in PA 135°. This object is bright enough to be seen with the 11×80 finder scope. The central core includes a stellar point in the middle. There are eight stars involved with faint spiral structure in the outer sections in this lovely galaxy. From the darkest sites, the spiral arms of this galaxy shimmer and sparkle with mottling.

(See Fig. 9.9)

a

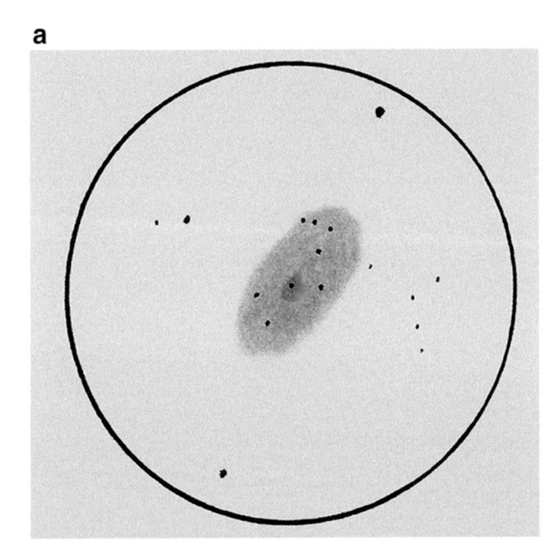

b

Fig. 9.9 (**a**) 13″ f/5.6; NGC 2403; FOV 25′; MAG 135×. (**b**) TEC 140 f-7 QHY8 camera. *Photo: George Kolb*

Object	NGC 3115
Other names	**H I 163**
Type	**GALAXY**
Mag	**8.9**
Size	**8.3′×3.2′**
Class	**E6**
Surface brightness	**11.9**
Constellation	**Sex**
RA	**10 05.2**
Dec	**−07 43**
Tirion	**13**
U2000	**279**
Description	**vB, L, vmE 46, vgsmbMEN**

6 in. f/8 S=7 T=7 Good site
90×—Bright, pretty large, very much elongated 3.5×1, much brighter middle, elongated nucleus. It is the Spindle Galaxy indeed. Averted vision makes it thicker.

13 in. f/5.6 S=6 T=8 Excellent site
11×80 finder—just seen as an elongated glow.
100×—Bright, large, very much elongated 4×1 in PA 45, suddenly much brighter in the middle.
220×—The center is a bright envelope that has an oval, very bright nucleus, all of which at elongated 3×1 in the same PA as the main body of the galaxy.

After spending time with low–surface-brightness detail in the galaxies up to now, this one has a high surface brightness so that you can see the difference (See Fig. 9.10).

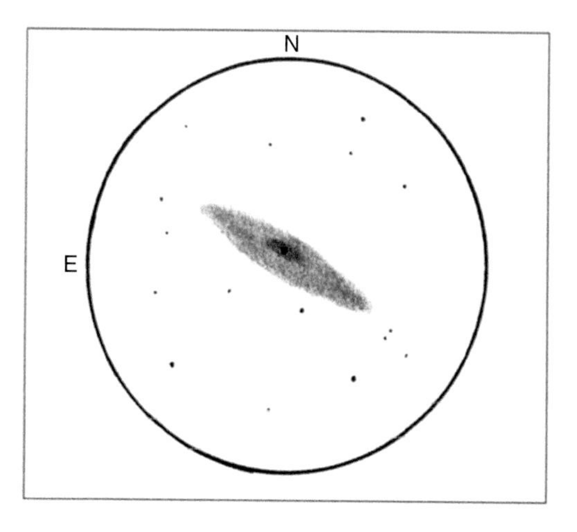

Fig. 9.10 13″ f/5.6; NGC 3115; FOV 15′; MAG 220×

Object	**NGC 4027**
Other names	**H II 296**
Type	**GALAXY**
Mag	**11.1**
Size	**3.0′ × 2.3′**
Class	**SB (s) dm**
Surface brightness	**13.2**
Constellation	**Crv**
RA	**11 59.6**
Dec	**–19 15**
Tirion	**13**
U2000	**327**
Description	**pF, pL, R, rr, st16**

6 in. f/8 S=7 T=7 Good site
90×—Pretty faint, pretty large, very little elongated 1.2×1, bright middle.

13 in. f/5.6 S=7 T=8 Excellent site
150×—Pretty faint, pretty large, very little elongated 1.2×1 in PA 120, much brighter middle. There is a faint star just to the northeast of the core.
220×—Arms are very mottled. This galaxy exhibits the bizarre effect of a bright area around the core that is elongated in PA 75, while the main galaxy is elongated in PA 120. I cannot think of another example of an object with a bright core that is skewed in position angle relative to its main body.

Look for that bizarre or unique observation; it can make for a grand and memorable evening. Notice the NGC description includes “rr”, meaning pretty well resolved. Obviously, no NGC observer really resolved this distant galaxy into individual stars, but I believe that they were referring to the mottling in the arms from H II regions (See Fig. 9.11).

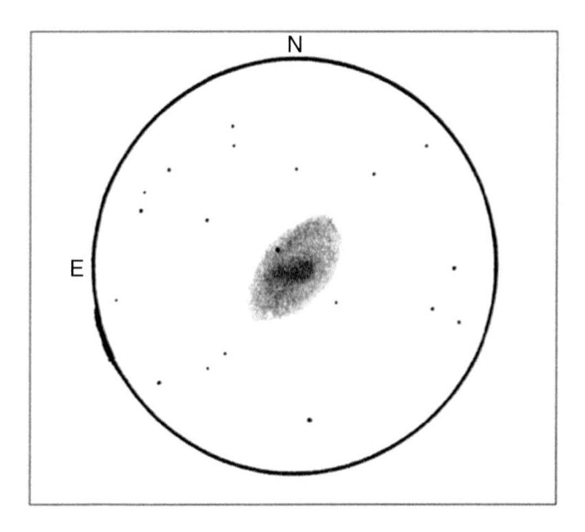

Fig. 9.11 13″ f/5.6; NGC 4027; FOV 15′; MAG 220×

Object	**NGC 4216**
Other names	**H I 35**
Type	**GALAXY**
Mag	**10.0**
Size	**8.5′×1.7′**
Class	**Sb**
Surface brightness	**12.8**
Constellation	**Vir**
RA	**12 15.9**
Dec	**+13 09**
Tirion	**13**
U2000	**193**
Description	**vB, vL, vmE17, sbMN**

6 in. f/8 S=7 T=8 Excellent site

90×—Bright, large, very much elongated 6×1, somewhat brighter middle with almost stellar nucleus. This excellent edge on galaxy is more prominent with averted vision.

13 in. f/5.6 S=7 T=8 Excellent site

100×—pretty bright, large, very much elongated 4×1 in PA 20, pretty suddenly much brighter middle. A nice edge–on, but what is appealing is the unique field of view. Within 30′ of NGC 4216 are two more edge-on galaxies. There are few other locations in the sky where three elongated galaxies are within 1° of one another. The companion to the southwest is pretty bright, pretty large, elongated 3×1 and very little brighter in the middle with a stellar nucleus. The northeast companion has a low surface brightness so it is faint, elongated 2.5×1 and pretty small; averted vision helps to pick it out of the background.

220×—Concentrating on NGC 4216, shows the central bright section is also elongated in the same PA as the main body of the galaxy. There is a pretty faint star to the east of the nucleus of 4216.

Sometimes the field of view is the reason that an object stands out and is worth a return visit. I will leave it you to find the designations of the two companions. Use that star chart (See Fig. 9.12).

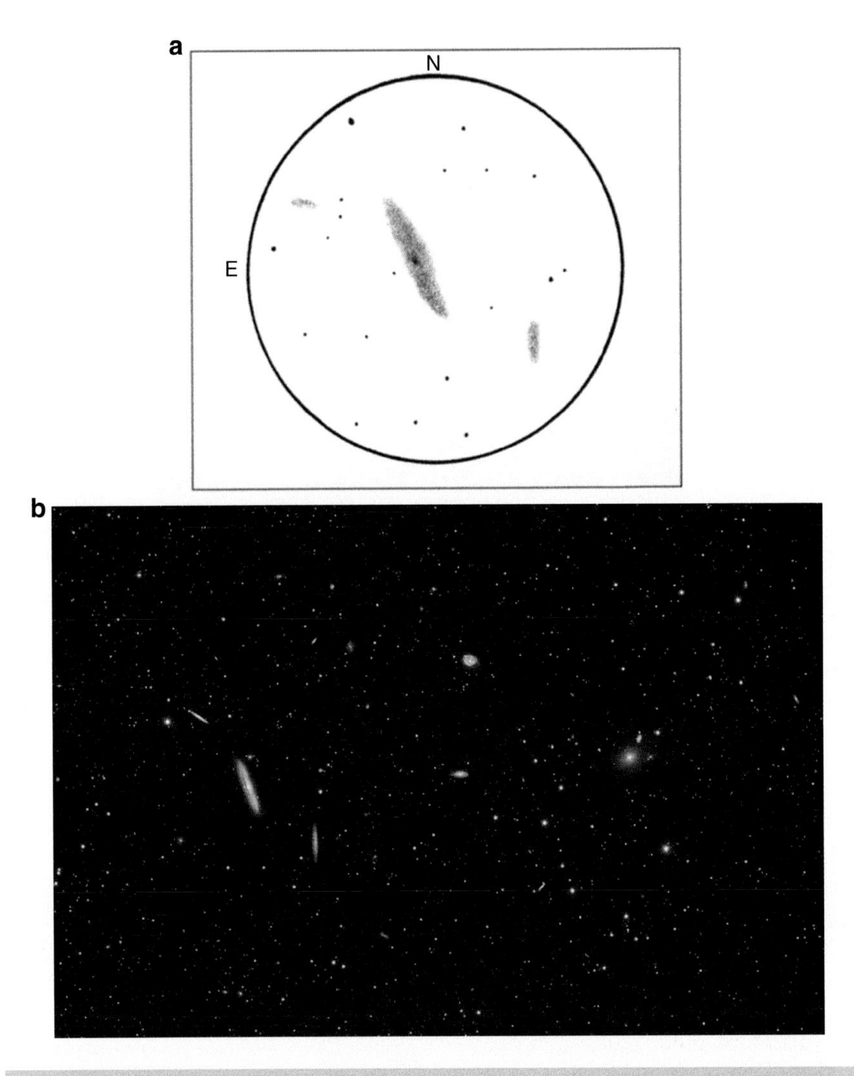

Fig. 9.12 **(a)** 13″ f/5.6; NGC 4216; FOV 30′; MAG 150×. **(b)** TEC 160 SBIG 8300. *Photo: Lee Buck*

Object	**NGC 4565**
Other names	**H V 24**
Type	**GALAXY**
Mag	**9.6**
Size	**15.5′×1.9′**
Class	**Sb**
Surface brightness	**13.3**
Constellation	**Com**
RA	**12 36.3**
Dec	**+26 00**
Tirion	**7**
U2000	**149**
Description	**!!B, eL, eE135, v sbMN=*10-11**

4 in. f/8 S=7 T=8 Excellent site
60×—Pretty bright, pretty large, elongated 3×1 in PA 135°, bright middle, averted vision makes it larger. I know that the dark lane is there, I just can't see it with a 4 in. telescope.

13 in. f/5.6 S=7 T=7 Excellent site
11×80 finder—just seen as faint and elongated
100×—Bright, very large, extremely elongated 7×1 in PA 135, suddenly brighter in the middle with an almost stellar nucleus. The dark lane is easy to see for about half the length of the galaxy and trails off at the very edges.
150×—This is the best view in the 13 in.; the dark lane shows scalloped detail with averted vision and the galaxy covers about 80 % of the field of view.

36 in. f/5 Texas Star Party 96 S=6 T=8 Excellent site
225× with 20 mm Nagler—In a mammoth telescope this galaxy covers the entire width of the field of view; averted vision makes it double in thickness—the dark lane is obvious and has much "scalloping" in several places with direct vision. The nuclear bulge is easy and is about twice the width of the rest of the galaxy. The core "sits" on the dark lane. Amazing view!

This spectacular edge-on galaxy is also a companion to Comet Coe. The story goes like this: years ago, I had just completed my first large aperture telescope, a 17.5 in. Dobsonian, and wanted to see what it could do under a really dark sky. I trucked it to a club gathering at one of our best sites. When I observed NGC4565 it had an obvious fuzzy, round companion located at the end of a curved chain of stars. I immediately thought that this was a comet. After showing it to A.J. and several other club members, I looked it up in *Burnham's Handbook*. The companion galaxy is quite obvious in a photograph there. Oh well, so much for fame and fortune. One of my "friends" pointed out that it could be a really long-period comet that is coming directly at the Earth (See Fig. 9.13).

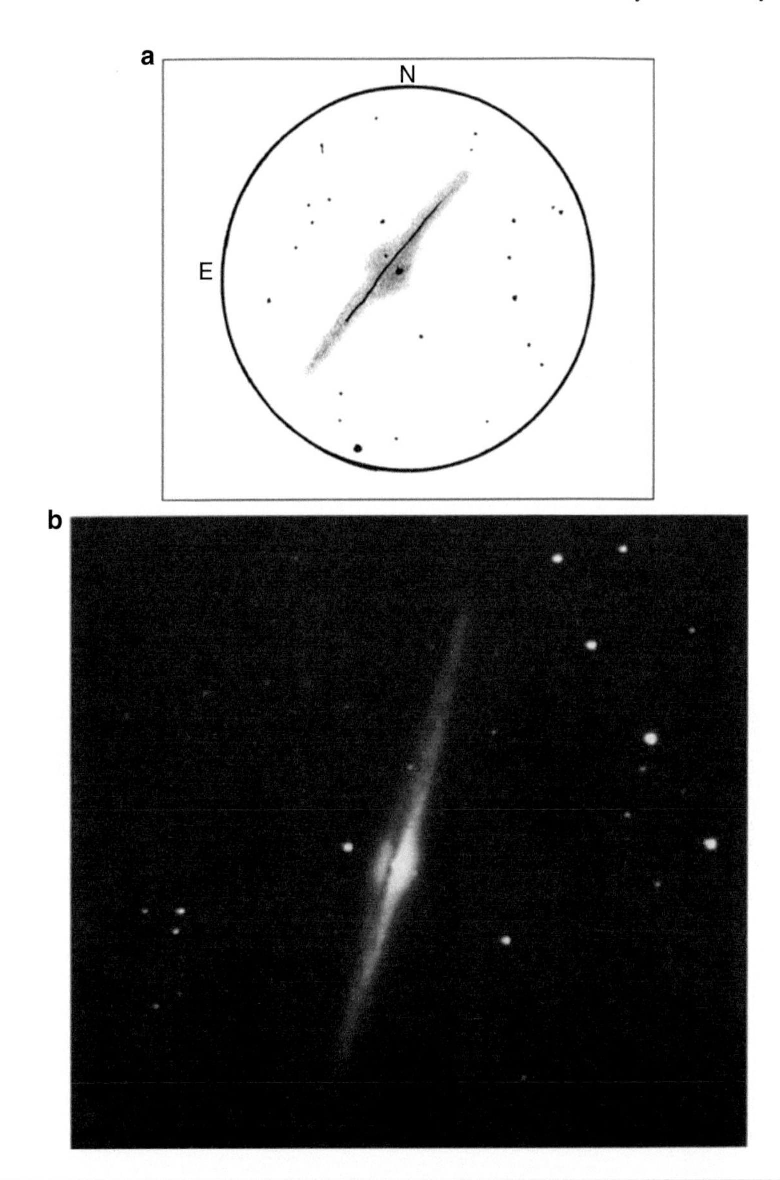

Fig. 9.13 (**a**) 3″ f/5.6; NGC 4565; FOV 25′; MAG 150×. (**b**) C-14 f/6; 1½ h exposure. *Photo: David Healey*

Object	**NGC 4567**
Other names	**H IV 8**
Type	**GALAXY**
Mag	**11.3**
Size	**3.0′ × 2.5′**
Class	**Sb**
Surface brightness	**13.1**
Constellation	**Vir**
RA	**12 36.6**
Dec	**+11 16**
Tirion	**13**
U2000	**194**
Description	**vF, L, np of Dneb**

6 in. f/8 S = 7 T = 8 Excellent site

90×—Faint, small, elongated 1.5 × 1, little brighter middle. The companion is even fainter.

13 in. f/5 S = 8 T = 9 Superior site

NGC 4567 and 4568 form the famous galaxy pair, the Siamese Twins.

150×—Both galaxies are pretty bright, pretty large, brighter in the middle and elongated 2 × 1. The two galaxies are joined on their eastern tips and the resulting "V" shape is obvious at this power with direct vision. The two galaxies are at an angle of approximately 30° to each other. NGC 4567 is elongated in PA 90, 4568 is elongated in PA 60. The northern galaxy of the pair is NGC 4567; it is a little smaller and fainter than 4568.

220×—Too much power; some of the detail is lost by magnifying these galaxies until they appear out of focus or "mushy" (See Fig. 9.14).

On objects that you believe might show more detail on a terrific night, keep raising the magnification until no more can be seen.

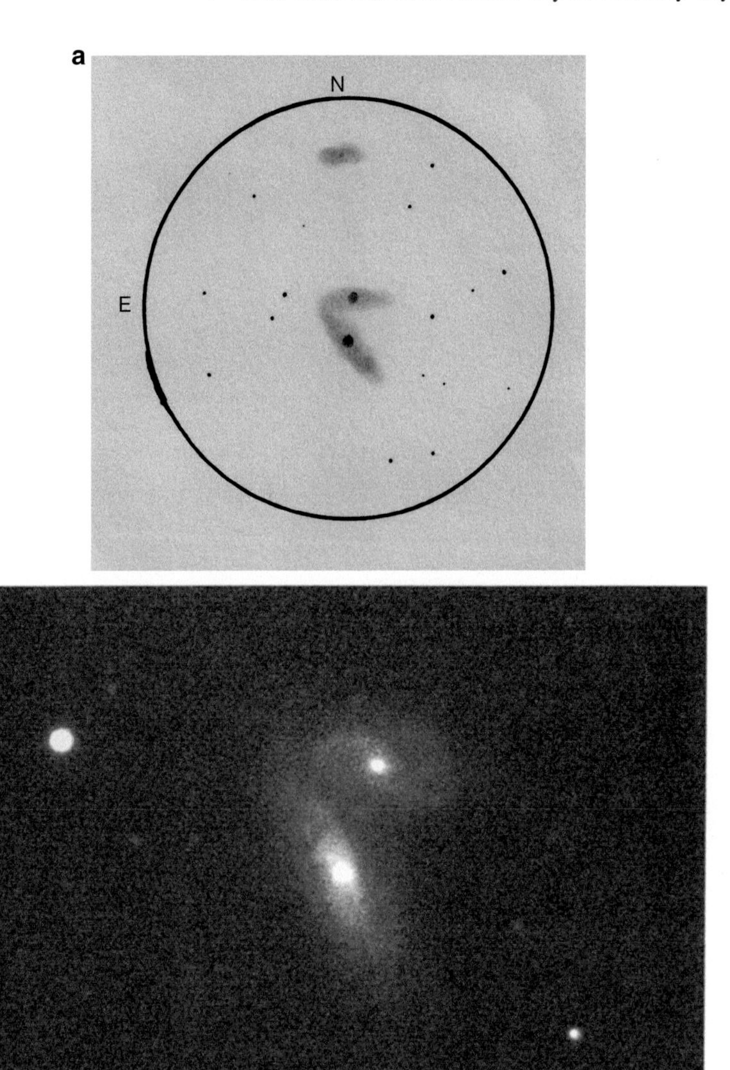

Fig. 9.14 (**a**) 13″ f/5.6; NGC 4567; FOV 25×; MAG 150×. (**b**) 12″ L×200 at f/7; ST 7E CCD; 6 min exposure. *Photo: Larry E. Robinson*

Object	**NGC 5128**
Other names	**Dunlop 482**
Type	**GALAXY**
Mag	**6.8**
Size	**18.2′ × 14.5′**
Class	**S0p**
Surface brightness	**13.5**
Constellation	**Cen**
RA	**13 25.3**
Dec	**–43 01**
Tirion	**21**
U2000	**403**
Description	**!!vB, vL, vmE122, bifid 17**

6 in f/8 S = 7 T = 8 Excellent site

90×—Bright, large, elongated 1.8 × 1, very irregular figure with a bright middle and bifurcated by a dark lane. On an 8/10 night the lane is easy even at moderate power. There are two stars involved. Averted vision makes it much larger. A great view of a great object.

17.5 in. f/4.5 S = 7 T = 10 Superior site

100×—Bright, very large and round. The dark band across this galaxy is easy, even at this low power. There are three stars superimposed across the face of this object. Averted vision makes the size of the outer sections of this galaxy grow much larger.

150×—Now there is detail to see within the dark band. On the eastern end of the band is a faint, thin bright streak that is centered within the band. On the western end of the dark rift is an obvious star, held with direct vision. Another star is in the southern half of this bisected galaxy. A unique object.

Sometimes you must be patient; for northern observers this object only gets above the southern horizon during a short period of time in the spring, so be ready. This is one of the deep–sky objects that was on my observing list when we made that fateful trip to the MMT (See Fig. 9.15).

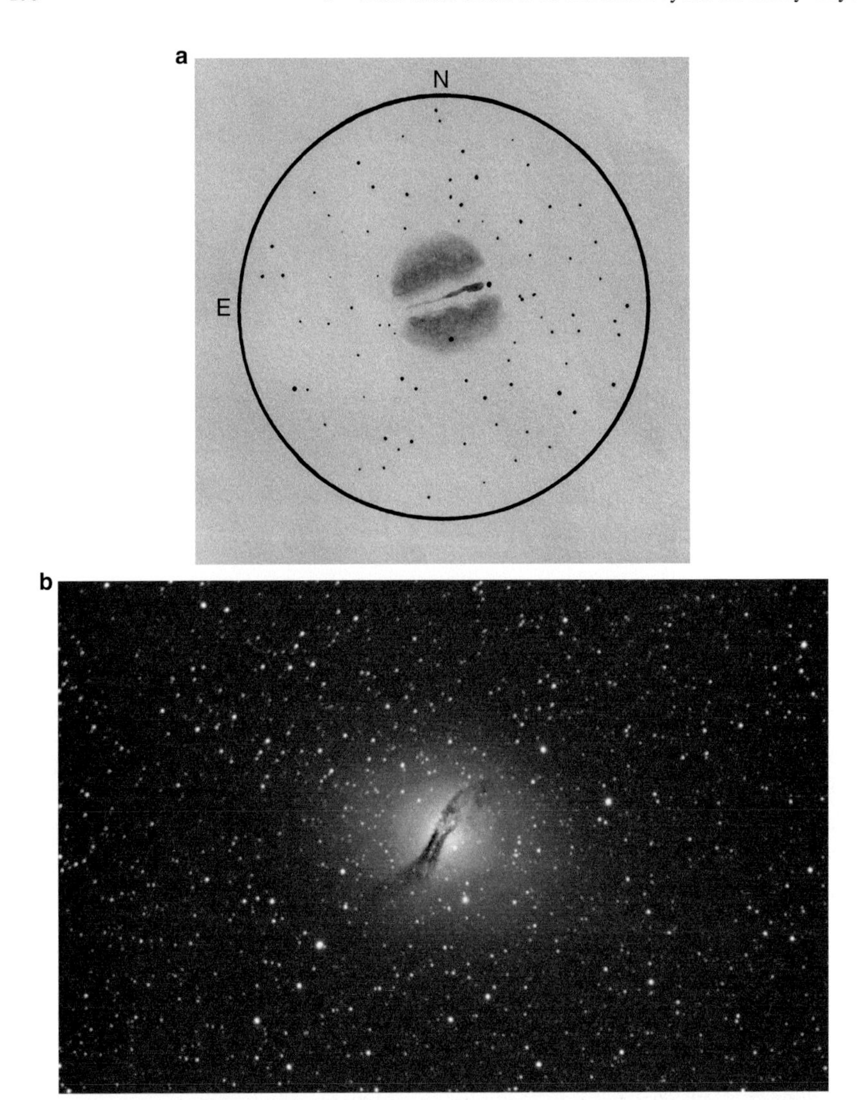

Fig. 9.15 (**a**) 13″ f/5.6; NGC 5128; FOV 25′; MAG 150×. (**b**) TEC 140 f-7 QHY8 camera. *Photo: George Kolb*

Object	**NGC 6946**
Other names	**H IV 76**
Type	**GALAXY**
Mag	**8.8**
Size	**14′**
Class	**Sc**
Surface brightness	**13.8**
Constellation	**Cep**

(continued)

(continued)

RA	**20 34.8**
Dec	**+60 09**
Tirion	**3**
U2000	**56**
Description	**vF, vL, vsbM, rr 17**

6 in. f/6 S=7 T=8 Excellent site

45×—Faint, pretty large, a little brighter in the middle and elongated 1.2×1 in PA 135. Raising the power to 100× shows some detail, mostly mottling in the outer sections. That is much more prominent with averted vision. At even higher powers this low surface brightness galaxy almost disappears, even on a night I rated 8/10 for transparency.

13 in. f/5.6 S=7 T=8 Excellent site

100×—Pretty bright, large, much brighter in the middle, elongated 1.8×1 in PA 30; it is irregularly elongated. There is some mottling in the arms. The core is elongated 2×1 in PA 30.

150×—A hint of spiral structure at this higher power. Next, I add the monk's hood for maximum contrast. The arms now sparkle with mottling and even with direct vision there is a curved section to the northeast that splits into two very faint spiral arms. Another stubby spiral arm feature is seen to the southwest. I try using the UHC nebula filter to look for H II regions in the arms, but none of the bright spots seem to respond to the filter.

220×—This high a power eliminates the contrast seen in the spiral arms, but the core is much easier and now contains a tiny, almost stellar nucleus. I estimate that the nucleus is about twice the size of the seeing disk of nearby stars.

25 in. f/6 S=6 T=8 Excellent site
McDonald Observatory

185×—This is easily my best view of this object. Three spiral arms are obvious with direct vision and there are 16 stars involved within the spiral arms. Several of these knots are much larger than the seeing disk and seem to be distant star clusters. This galaxy is in a rich Milky Way field.

Objects with low surface brightness respond to the atmosphere; observing them on a night with poor or mediocre transparency is not your best use of observing time. Observe some open clusters and high-surface-brightness objects while you wait for a better night (See Fig. 9.16).

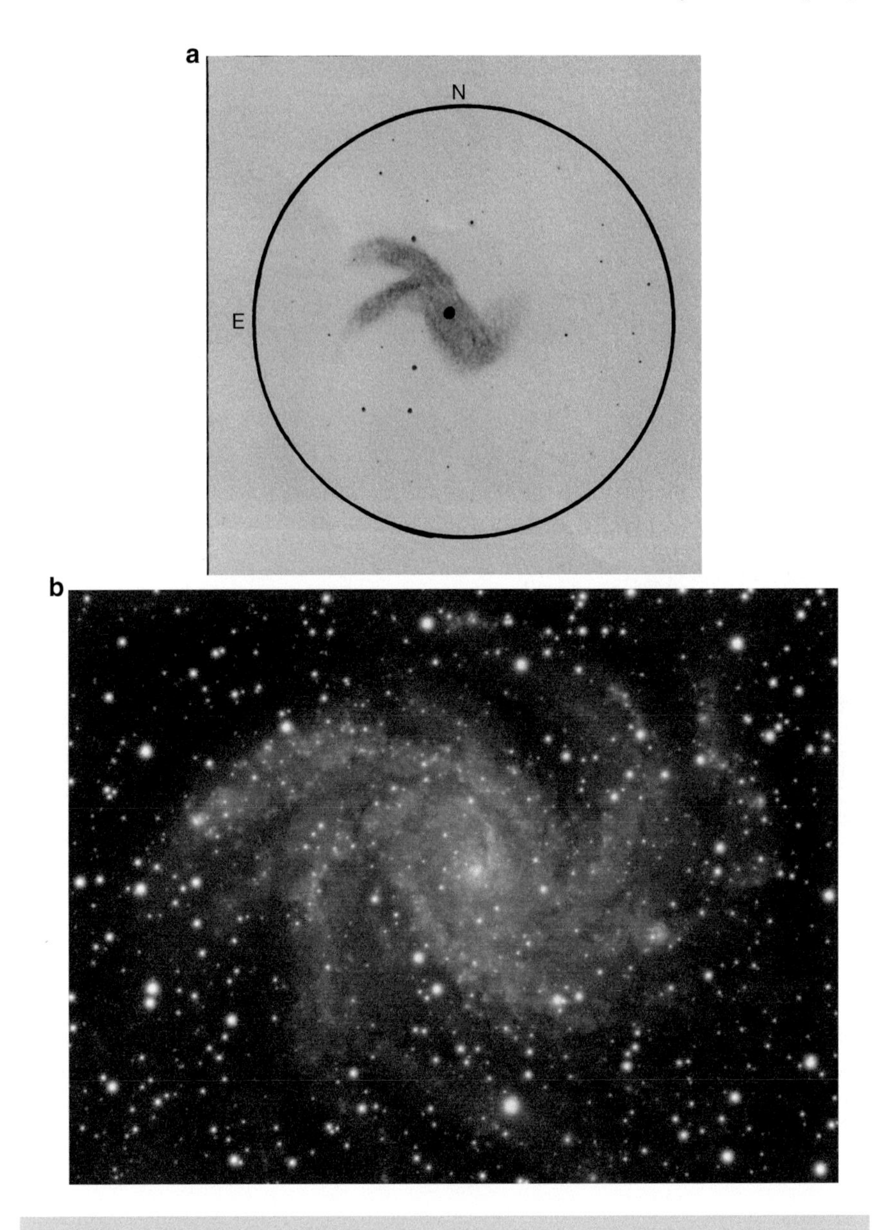

Fig. 9.16 (**a**) 13″ f/5.6; NGC 6946; FOV 20′; MAG 150×. (**b**) C14 at f7,6 Atik 314L+ camera. *Photo: Parijat Singh*

Object	NGC 7201
Other names	
Type	**GALAXY**
Mag	**12.8**
Size	**1.7′ × 0.7′**
Surface brightness	**12.4**
Constellation	**PsA**
RA	**22 06.5**
Dec	**–31 16**
Tirion	**23**
U2000	**383**
Description	**F, R, gbM, 1st of 4**

13 in. f/5.6 S = 6 T = 8 Good site

150×—Faint, pretty large, bright middle with an almost stellar nucleus. It is elongated 2 × 1 in PA 120. Averted vision makes this galaxy grow in size. Notice the NGC description: it says 1st of 4, which means that there are other objects in this same field of view. A look at the image in the *Vicker's CCD Atlas* shows me that NGC 7202 is simply a very faint star. So, let's see what other galaxies are in the small group.

NGC 7203 at 150×—Faint, pretty small, little brighter middle and a little elongated 1.2 × 1 in PA 60.

NGC 7204 at 150×—Pretty faint, pretty large, elongated 1.8 × 1 in PA 60, brighter middle; averted vision makes it grow in size.

In between 7203 and 7204 is another galaxy: MCG–5-52-28. I never held it steady; it is a very difficult object in the 13 in. scope. The best I could do is hold it about 40 % of the time with averted vision. If you have a scope in the same size range as mine and are looking to determine your limits, this tough galaxy can provide a good test.

This is a pretty faint, but interesting, galaxy grouping. So, don't be concerned about looking away from the well-worn path. Lots of challenging objects and interesting fields of view exist in constellations that don't get lots of magazine coverage (See Fig. 9.17).

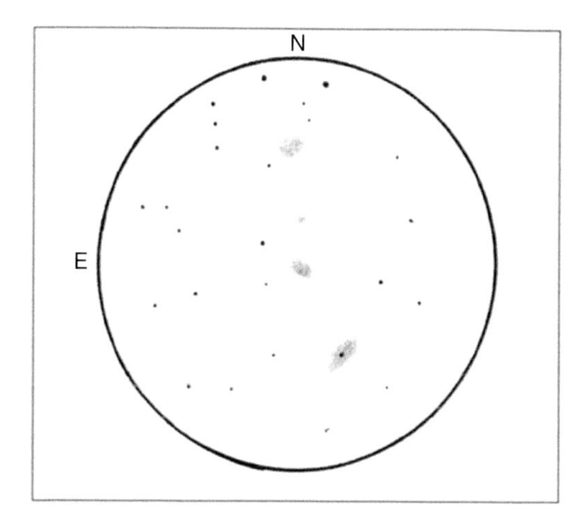

Fig. 9.17 13″ f/5.6; NGC 7201; FOV 25′; MAG 150×

Object	NGC 7479
Other names	H I 55
Type	GALAXY
Mag	10.9
Size	4.4′ × 3.4′
Class	SBb
Surface brightness	13.4
Constellation	Peg
RA	23 04.9
Dec	+12 19
Tirion	17
U2000	213
Description	pB, cL, mE12, bet 2 st 17

6 in. f/8 S = 7 T = 7 Good site

60×—Pretty faint, pretty large, elongated 1.8 × 1 in PA 15, very little brighter middle. It is located between two stars. Using averted vision makes it thicker.

17.5 in. f/4.5 S = 6 T = 7 Good site

165×—Pretty bright, large, much brighter middle and elongated 4 × 1 in PA 20. This object is a very nice barred spiral and that structure can be seen on good evenings. The bar is about 5′ in length and each end has a curved glow attached. It looks like a two-armed garden sprinkler in action. Averted vision makes the galaxy grow in size (See Fig. 9.18).

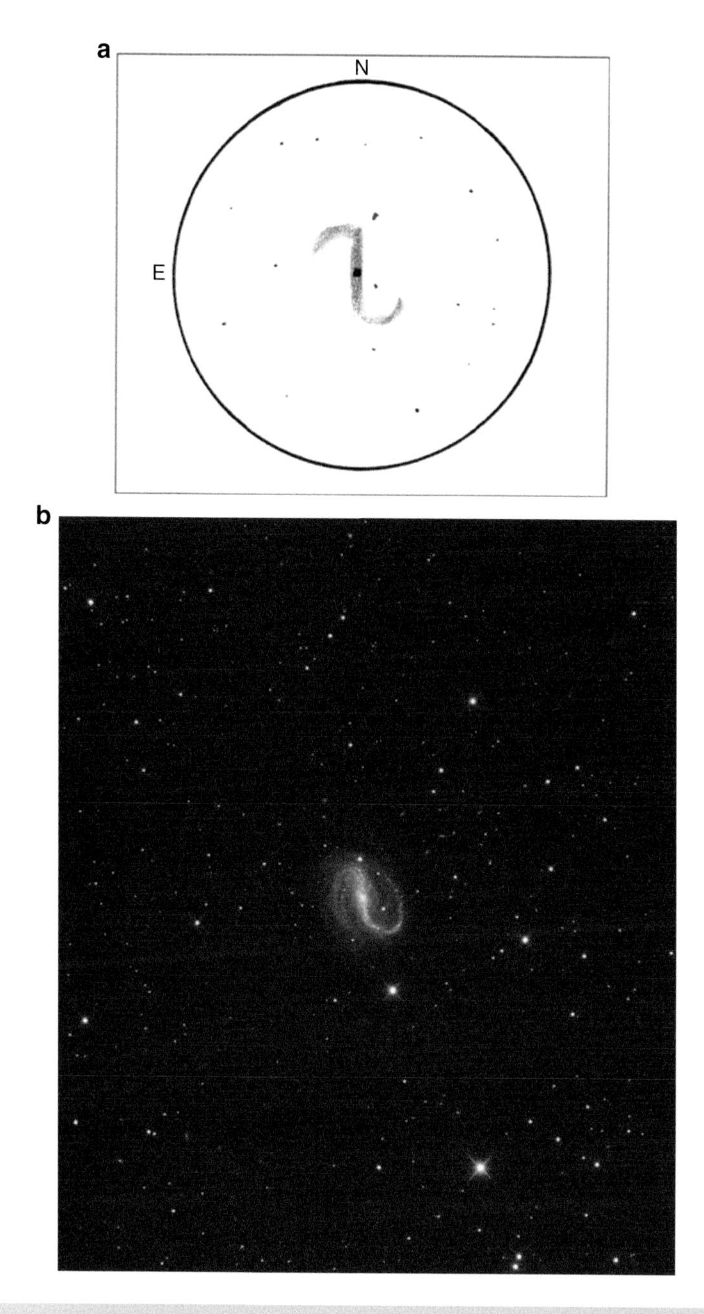

Fig. 9.18 (**a**). 13″ f/5.6; NGC 7479; FOV 20′; MAG 150×. (**b**) AT 10 RC QSI 640wsg camera. *Photo: John Dwyer*

Chapter 10

What Can I Observe in a Cluster of Galaxies?

George Abell has passed away but one of the most enduring legacies he has left astronomy is a catalog of galaxy groups that he compiled in the 1950s. Abell examined the Palomar Sky Survey plates to find clusters of galaxies. Values were assigned according to the richness of the number of galaxies and the distance of the cluster. He found that the magnitude of the tenth-brightest galaxy was a good indicator of the relative brightness of the cluster. This groundbreaking work has stood the test of time as a valuable system for evaluating galaxy clusters. Table 10.1 contains information on the 32 closest Abell galaxy groups, the ones assigned distance values of "0" or "1". Brian Skiff has observed many of these clusters over the years and has informed me that this is a very reasonable criterion for making a list of the best and brightest galaxy clusters. Brian is very knowledgeable and so it is this criterion I chose for the listing in the table.

The magnitude column is for the tenth-brightest galaxy in that group.

The *Uranometria* chart number column will contain several chart numbers if the area of the sky overlaps more than one chart. The size listing is in either arc minutes or square degrees, depending on which type of information I could find. Recently, a new version of Uranometria has been released but I left in the older page references because few people I know have purchased the newer release. If you do need to update my information, I have included the RA and Dec, so you can find the pages.

My observations of some of these galaxy clusters are included. There is also a photo reference if a picture of this cluster has been in *Sky and Telescope, Deep Sky* or *Astronomy* magazine. If the cluster is covered in the *Observer's Handbook* by Luginbuhl and Skiff then there is a reference to a page number.

S.R. Coe, *Deep Sky Observing*, The Patrick Moore Practical Astronomy Series,
DOI 10.1007/978-3-319-22530-2_10

Table 10.1 The nearest Abell galaxy clusters

Abell no.	Constellation	RA right ascension (2000)	Declination	Magnitude	Uranometria 2000	Size	Notes
194	Psc	01 25.5	–01 22	13.0	218	30′	Includes NGC 541 and Minkowski's object, Arp 133
262	And	01 52.7	+36 09	13.0	92	120′	Includes NGC 708 and 753; rather loose and irregular.
347	And	02 25.1	+41 48	13.0	62	40′	0.5° south preceding NGC 891.
400	Cet	02 57.6	+06 02	13.9	175/176	3.7°	17 Gal/°.
407	Per	03 01.8	+35 51	14.7	93/94	1.8°	22 Gal/°.
426	Per	03 19.7	+41 30	13.0	63	30′	Centered on NGC 1275; Milky Way makes ID difficult. Photo: *Sky and Telescope,* Jan. 1989, p. 20. Luginbuhl and Skiff, p. 192.
539	Ori	05 16.7	+06 28	14.4	180	2.4°	27 Gal/° Photo: *Sky and Telescope,* Dec. 1989, p. 670
548	Lep	05 47.1	–25 38	13.7	316	4.5°	14 Gal/°.
569	Lyn	07 09.2	+48 38	13.8	68	4.1°	9 Gal/°.
576	Lyn	07 21.4	+55 44	14.4	42	2.4°	27 Gal/°.
779	Lyn	09 19.9	+33 46	13.8	103	4.1°	9 Gal/° Luginbuhl and Skiff, p. 162.
1060	Hyd	10 36.8	–27 32	12.7	325/366	12°	5 Gal/° Luginbuhl and Skiff, p. 133. Photo: *Sky and Telescope,* Dec. 1976, p. 430.
1185	UMa	11 10.6	+28 46	14.0	106/146	40′	Incl NGC 3550 and Ambartsumian's Knot; a dwarf at end of plume. Photo: *Sky and Telescope,* Jan. 1988, p. 20.
1213	UMa	11 16.4	+29 17	14.5	106	2.2°	30 Gal/°.
1228	UMa	11 21.5	+34 20	13.8	106	4.1°	15 Gal/°.
1314	UMa	11 34.8	+49 03	13.9	47/73/74	3.7°	10 Gal/°.
1367	Leo	11 44.5	+19 50	14.0	147	30′	More GALAXYS > 14 mag than any GALCL. Photo: *Deep Sky,* no. 10, Spring 1985, p. 10, Luginbuhl and Skiff p. 147.

1377	UMa	11 45.6	+55 53	14.0	47	30′	At limit of 16″; no NGC or IC members; * 6 mag superimp.
1656	Com	12 59.8	+27 59	11.0	108/149	120′	Dense GALCL for amateurs; 72 brighter than 15 mag in 2°. Photo: *Deep Sky,* no. 10, Spring 1985, p 8. Finder chart in Luginbuhl and Skiff, p. 88.
1736	Hyd	13 26.8	−27 08	14.8	330/370	1.7°	24 Gal/°
2065	CrB	15 22.1	+27 39	14.0	154	30′	Brightest 6 Gal 15.5 mag, 40 in 1/2° field to 17 mag Photo: *Sky and Telescope,* May 1990, p. 563. Good finder charts and info. on this very distant cluster.
2147	Her	16 02.2	+15 55	13.8	200	4.1°	15 Gal/°
2151	Her	16 05.1	+17 43	15.0	155	40′	20 galaxies 14–15 mag; Hercules Galaxy Cluster. Photo: *Sky and Telescope,* Jan. 1988, p. 20.
2152	Her	16 05.3	+16 27	13.8	155/200	4.1°	15 Gal/°
2162	CrB	16 12.5	+29 32	13.7	113	4.5°	8 Gal/°
2197	Her	16 27.7	+40 55	14.0	80	60′	Centered on E–W chain of NGC 6146, 6160 and 6173.
2199	Her	16 28.6	+39 31	13.0	80/114	40′	Centered on NGC 6166. Photo: *Sky and Telescope,* Jan. 1988, p. 17.
2634	Peg	23 38.3	+27 03	13.8	89/124	4.1°	15 Gal/°
2666	Peg	23 50.9	+27 10	13.8	89/125	4.1°	9 Gal/°
3526	Cen	12 48.9	−41 02	13.2	402	2.1°	Centaurus I, a 2° long chain.
3565	Cen	13 36.7	−33 58	14.0	370	2.0°	IC 4296 group; not very condensed.
3574	Cen	13 49.2	−30 17	13.4	371	1.5°	IC 4329 group; bright members but somewhat sparse. Photo: *Deep Sky,* Spring 1986, p. 22.

I have also included two other groupings of galaxies that did not receive an Abell Galaxy Cluster (AGC) number. That is because there are not enough galaxies to constitute a rich cluster. However, these galaxy groups can provide interesting views of a part of the sky that contains several non -stellar objects in one field of view. I find that type of field fascinating. You can start to find this type of field for yourself; just look for overlapping galaxy symbols on your star chart. The two galaxy groups from my notes are: the NGC 474 Group in Pisces and the famous Stephan's Quintet in Pegasus.

Some of the objects in these clusters are going to be a test for a large amateur telescope even on the best of evenings. So, save the galaxy clusters for nights with excellent transparency at a dark site.

NGC 474 Group in Pisces

13 in.	f/5.6	S=7 T=8	Excellent site

NGC 467 pretty faint, pretty small, not much brighter in the middle, little elongated 1.5×1 in PA 135 at 100×.

NGC 470 pretty bright, pretty large, not brighter in the middle, round, very mottled at 100×.

NGC 474 pretty bright, pretty large, much brighter in the middle, round at 100× the brightest of a group of three with NGC 467 and 470 (See Fig. 10.1).

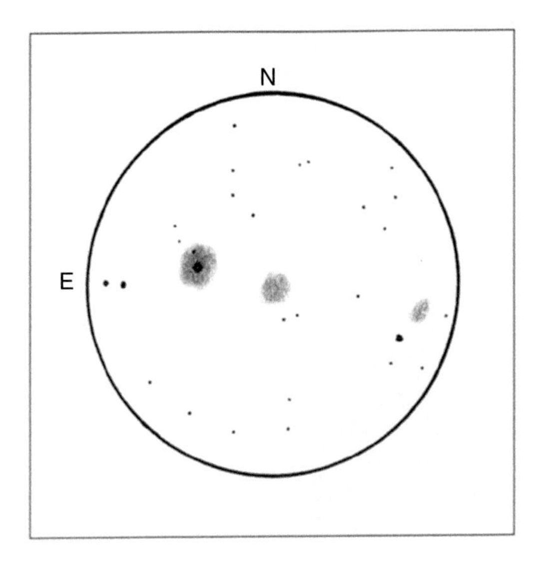

Fig. 10.1 13″ f/5.6; NGC 470; FOV 30′; MAG 100×

Abell 262

13 in.	f/5.6	S=8 T=9	Excellent site

135×—There are two centers to this galaxy cluster. One is around NGC 708; it is pretty faint and round. There are three other galaxies around NGC 708, all very faint; one is round, two are elongated. The other center of Abell 262 is NGC 785; it is pretty faint, round and brighter in the middle. It is surrounded by four fainter galaxies (See Fig. 10.2).

36 in.	f/5	S=7 T=9	Excellent site

200×—An absolutely gorgeous rich field of galaxies. NGC 708 is surrounded by 40 other galaxies within the field of view. Most are small and quite faint, but seeing so many galaxies in one view is a rare treat. Moving the large Dobsonian around two fields of view in each direction from NGC 708, I quickly count 100 other galaxies in this cluster.

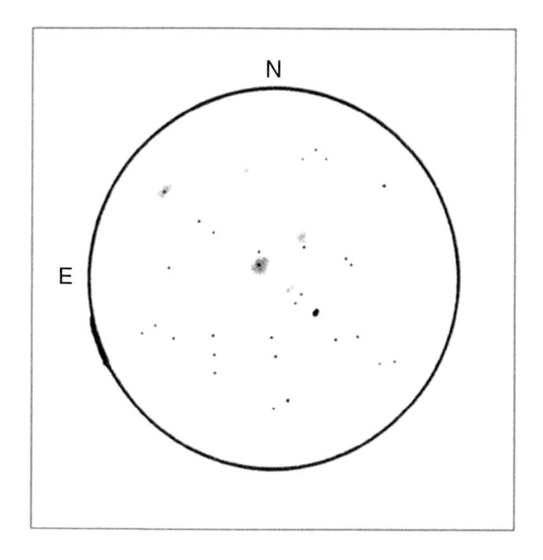

Fig. 10.2 13″ f/5.6; Abell 262; FOV 25′; MAG 150×

Abell 426 (also known as Perseus I)

17.5 in.	f/4.5	S=8 T=8	Excellent site

NGC 1275 is the central galaxy in the Perseus I cluster of galaxies.

165×—Pretty faint, small and little elongated is my observation of 1275. With averted vision I could pick out six other galaxies within 1° of NGC 1275. This is a pretty rich Milky Way field and that makes deciding what is a galaxy and what is a star quite difficult at times. I have only reported the objects I could definitely identify as galaxies (See Fig. 10.3).

36 in. f/5 S=6 T=8 Excellent site

200×—There are 17 bright galaxies and 58 other members within one field of view of the central section in this rich galaxy cluster. A few show some faint spiral structure, but most are elliptical or round. Tom Clark is the owner of this telescope. He points out what he calls "The River of Galaxies", a chain of galaxies off to the west of NGC 1275. It is a fascinating meander of little galaxies to follow out of the field of view.

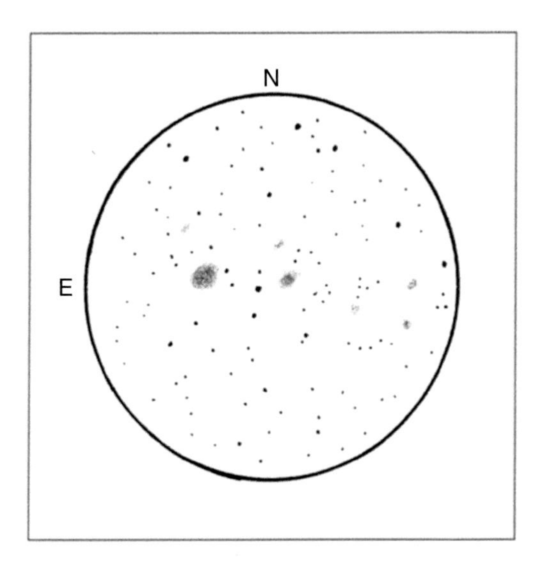

Fig. 10.3 17″ f/4.5; Abell 426; FOV 20′; MAG 165×

Abell 1060, the Hydra I Galaxy Cluster

13 in. f/5.6 S=8 T=9 Excellent site

All observations are at 165×.

NGC 3285 faint, pretty small, somewhat brighter in the middle, somewhat elongated; averted vision helps.

MCG 04-25-026 very faint, somewhat elongated, brighter middle.

MCG 04-25-025 pretty faint, little elongated, brighter middle.

NGC 3305 faint, small, round, small nucleus.

NGC 3308 pretty faint, somewhat elongated, slightly brighter middle.

NGC 3309 pretty faint, small, round, pretty bright star following.

NGC 3311 pretty faint, pretty small, somewhat brighter middle.

NGC 3312 pretty faint, pretty small, somewhat brighter in the middle, looks like

NGC 3311 with a brighter middle.

NGC 3314 very faint, elongated, not brighter in the middle, seen with averted vision only.

(continued)

(continued)

NGC 3315 pretty faint, round, small, somewhat brighter middle.
NGC 3316 very faint, very small, round, somewhat brighter middle.
MCG 04-25-050 extremely faint, small, not brighter middle, averted vision only.
IC 2597 extremely faint, pretty small, low surface brightness, very difficult, flickers in and out of view.
MCG-04-25-052 extremely faint, small, round, not brighter middle, averted vision only (See Fig. 10.4).

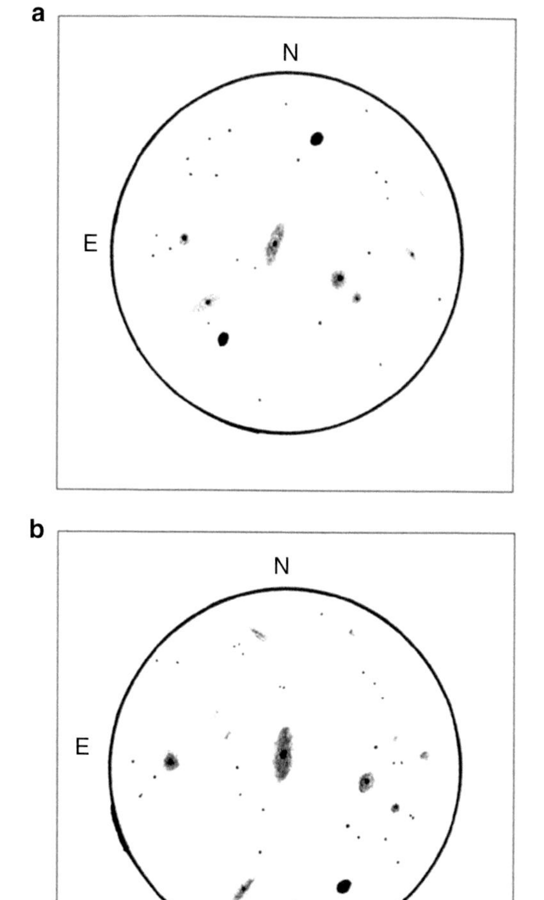

Fig. 10.4 (**a**) 13″ f/5.6; Abell 1060; FOV 30′; MAG 100×. (**b**) 36″ f/5; Abell 1060; FOV 25′; MAG 165×

Abell 1367

This rich galaxy cluster in Leo contains some of the most distant galaxies which can be seen in a modest telescope. In all the observations I will discuss, there is a background of unresolved galaxies, regardless of magnification—a view which forces the observer to contemplate the vastness of the Universe.

13 in. f/5.6 S=9 T=10 Superior site

165×—NGC 3842 is the central, pretty faint, galaxy. It is quite easily detected and it has four other galaxies within a 1° field of view.

220×—Going to higher power and adding the monk's hood to block off stray light, I could detect five other very faint galaxies. This is at my best location 120 miles from Phoenix, in mountains with very dark skies.

20 in. f/5 S=9 T=10 Superior site

180×—On the same night as above, Pierre Schwaar's big scope will allow me to count 22 galaxies. Most could not be classified as "easy", but they could be pointed out to friends, when they were at the eyepiece. Also, the 20 in. showed some detail within the galaxies that could not be seen at any power in the 13 in. As we used to say when I was racing cars, "There's no substitute for cubic inches."

36 in. f/5 S=6 T=8 Excellent site

200×—Moving the scope two fields of view about the central section, I can count 122 galaxies in this group. There are a wide variety of shapes and sizes. Spending some time with the central field, I can pick out 15 galaxies that I have never seen before. Easily my best view of this rich galaxy field (See Fig. 10.5).

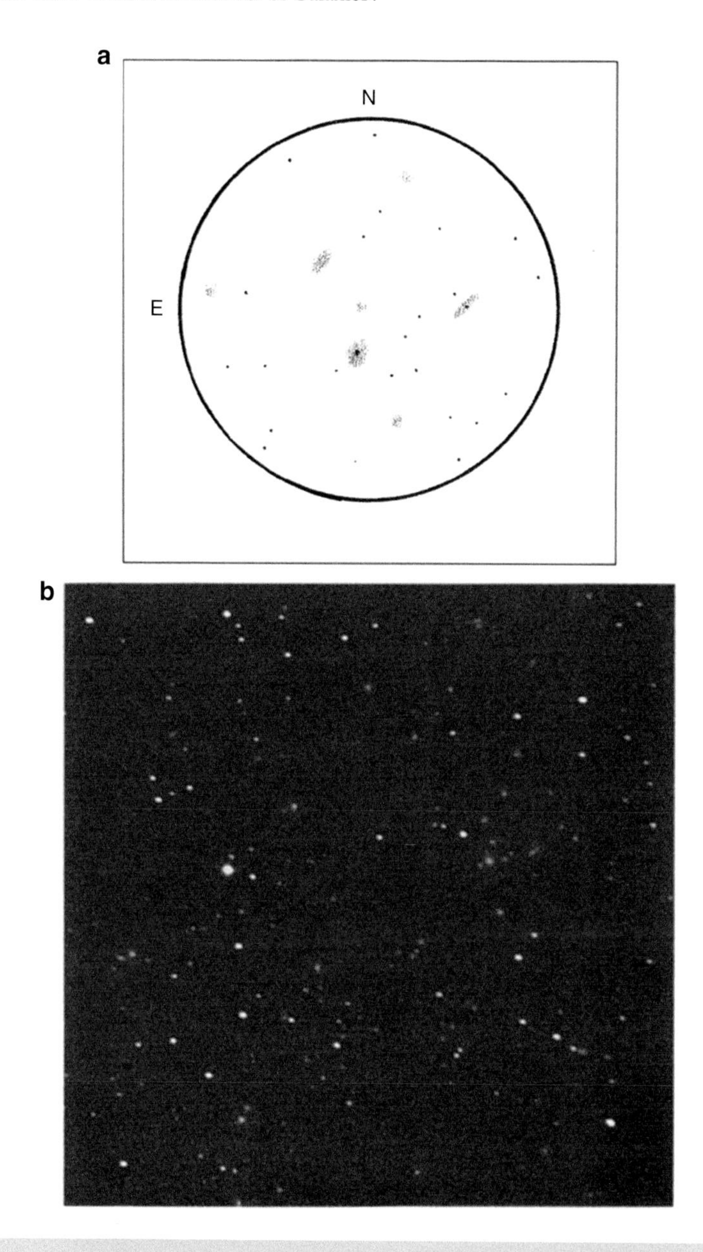

Fig. 10.5 (**a**) 13″ f/5.6; Abell 1367; FOV 20′; MAG 220×. (**b**) 14″ f/3.8; 45 min exposure. *Photo: Pierre Schwaar*

Abell 1656

Dr Fritz Zwicky surveyed this cluster and identified 804 galaxies brighter than 16.5 magnitude, so don't worry about running out of goodies to observe. A spectacular region that is also "lumpy" from the overwhelming background of galaxies.

13 in. f/5.6 S=8 T=9 Superior site

220×—NGC 4889 and 4874 are the center of this rich cluster. Both are pretty bright, pretty small and roundish. NGC 4889 has a bright core that makes it stand out brighter than any other cluster member. It is also surrounded by a swarm of very small, very faint galaxies. Using the dark hood I can see a total of five other members that are pretty faint and 20 very faint members. This count includes galaxies within two fields of view of the central pair of galaxies. If you push the scope about five fields of view away from the center, then you will see that the background gets darker. This is because the unresolved galaxies within this cluster are brightening up the field of view and creating a background glow.

36 in. f/5 S=7 T=8 Excellent site

200×—The central section around 4889 shows 42 galaxies within 20 arcmin. Another 126 galaxies are counted within one field of the central portion. The brightest 20 galaxies show detail: dark lanes, spiral arms, much brighter cores. Every shape, size and tilt that galaxies can display is seen by just maneuvering the scope about. Averted vision will show hundreds more galaxies at the limit of the big scope. WOW! This is the exclamation as people climb the ladder to view this lovely group with Larry Mitchell's 36 in. at the Texas Star Party. Those waiting at the foot of the ladder were patient enough to allow me to record the observation above. Across the dark and quiet observing field, I could hear people say, "You have gotto go see the Coma Galaxy Cluster in the 36 in (See Fig. 10.6)."

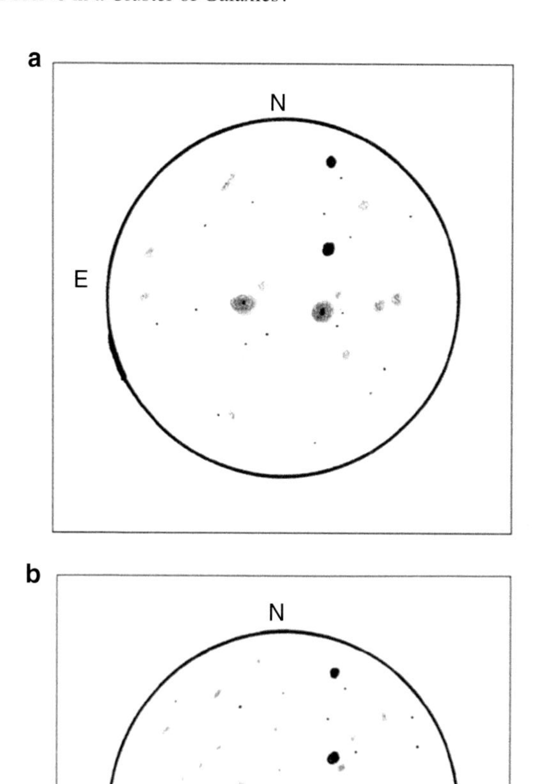

Fig. 10.6 (**a**) 13″ f/5.6; Abell 1656; FOV 20′; MAG 220×. (**b**) 36″ f/5; Abell 1656; FOV 20′; MAG 200×

Abell 2065

Estimates of the distance of this faraway galaxy cluster fall around a 1000 million light years, approximately the age of the rock strata at the bottom of the Grand Canyon. This grouping is about the same size and density as the Virgo Galaxy Cluster; it just happens to be about 17 times farther away.

13 in. f/5.6 S=7 T=9 Excellent site

220×—There are no individual galaxies that I can pick out, even on this terrific night. There is a very faint glow in the correct location, but no galaxies are seen with direct or averted vision, with or without the monk's hood.

20 in. f/5 S=9 T=1 Superior site

180×—There are two member galaxies that are very faint and another four galaxies that were suspected with averted vision. The field of view is a faint, mottled glow from galaxies just as the limit of detection.

36 in. f/5 S=6 T=8 Excellent site

200×—The view itself is not spectacular, but realizing what you are seeing is amazing. 11 small galaxies seen, all faint. Two are brighter than the rest, but all are not easy. The fainter galaxies come and go with the seeing. No real detail to be seen; one or two have a somewhat brighter middle, but not much. I am reminded of *The Hitchhiker's Guide to the Galaxy;* "Space is really, really BIG, you just can't picture how really big it really is (See Fig. 10.7)."

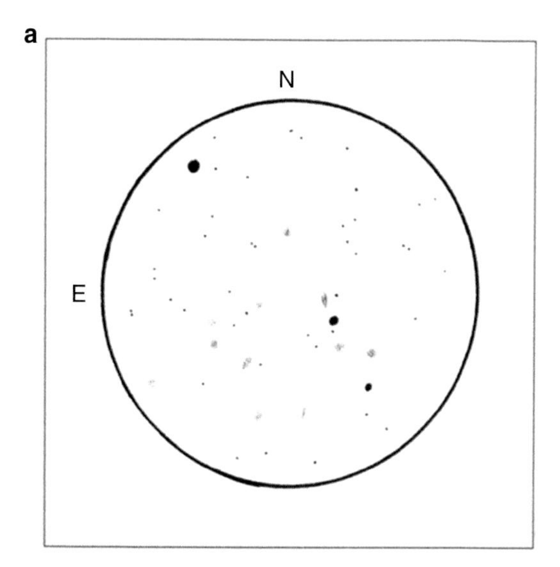

Fig. 10.7 (**a**) 36″ f/5; Abell 2065; FOV 20′; MAG 200×. (**b**) 20″ f/5; 1 h exposure. *Photo: Pierre Schwaar*

Abell 2151

13 in. f/5.6 S=8 T=10 Superior site

150×—Four round galaxies and one that is somewhat elongated are held with direct vision. The field has that "lumpy "aspect that means there are lots of unresolved galaxies in the eyepiece.

220×—Going to higher power and using the monk's hood, I can see another seven galaxies that are detected with averted vision. All are extremely faint and small.

36 in. f/5 S=7 T=8 Excellent site

320×—21 galaxies seen with direct vision in the central section of this galaxy cluster. That number is doubled with averted vision.

By moving the scope around another 40 galaxies can be seen within two fields of view of the central section. Very difficult to draw; even a dim red light wipes out many of the fainter galaxies. Just have to remember where they were and add them into the drawing (See Fig. 10.8).

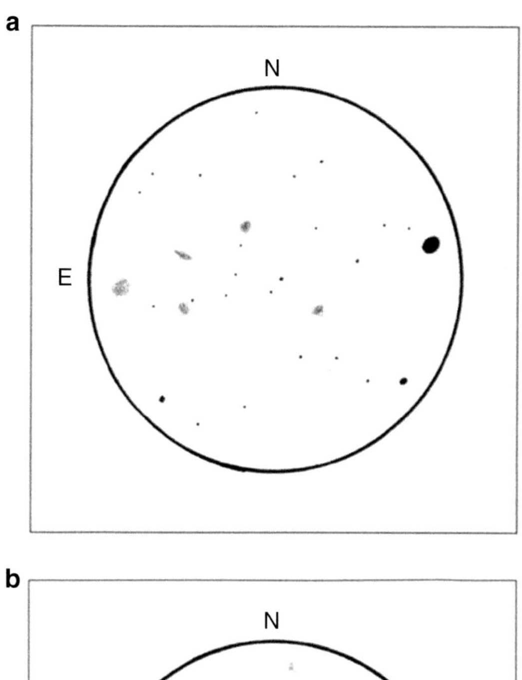

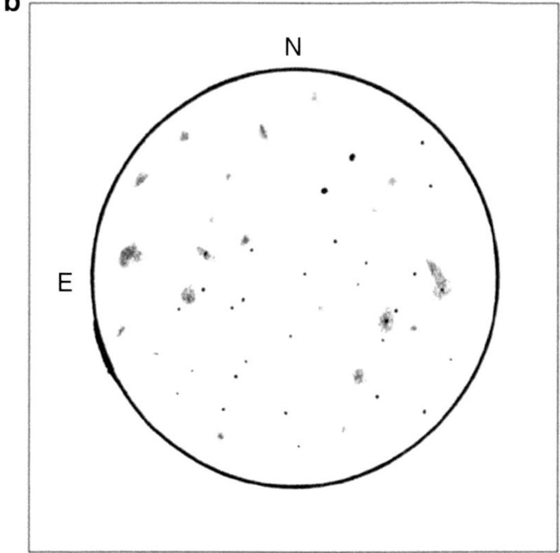

Fig. 10.8 (**a**) 13″ f/5.6; Abell 2151; FOV 25′; MAG 150×. (**b**) 13″ f/5; Abell 2151; FOV 15′; MAG 320×

Stephan's Quintet

This is easily the best known of the galaxy groupings around the sky. I am certain that many an amateur astronomer with a new, larger telescope has put Stephan's Quintet on his observing list. Here are some data on the five members of this group:

Object	**NGC 7317**
Type	**GALAXY**
Mag	**13.6**
Size	**0.7′ × 0.5′**
Surface brightness	**13.8**
Constellation	**Peg**
RA	**22 35.9**
Dec	**+33 57**
Tirion	**9**
U2000	**123**
Description	**vF, vS**

Object	**NGC 7318A**
Other names	**UGC 12099**
Type	**GALAXY**
Mag	**13.4**
Size	**1.7′ × 1.2′**
Surface brightness	**13.0**
Constellation	**Peg**
RA	**22 36.0**
Dec	**+33 58**
Tirion	**9**
U2000	**123**
Description	**eF, eS**
Notes	**7318A projects onto arm of 7318B**

Object	**NGC 7318B**
Other names	**UGC 12100**
Type	**GALAXY**
Mag	**13.1**
Size	**1.9′ × 1.2′**
Surface brightness	**13.9**
Constellation	**Peg**
RA	**22 35.9**
Dec	**+33 58**
Tirion	**9**
U2000	**123**
Notes	**7318B is distorted, colliding with 7318A**

Object	NGC 7319
Other names	UGC 12102
Type	GALAXY
Mag	13.1
Size	1.6′ × 1.2′
Surface brightness	13.8
Constellation	Peg
RA	22 36.1
Dec	+33 59
Tirion	9
U2000	123
Description	eF, eS
Notes	Distorted

Object	NGC 7320
Other names	UGC 12101
Type	GALAXY
Mag	12.6
Size	1.9′ × 1.0′
Surface brightness	13.5
Constellation	Peg
RA	22 36.1
Dec	+33 57
Tirion	9
U2000	123
Description	F, vS
Notes	PA 132, brightest in Stephan's Quintet

13 in. f/5.6 S=7 T=8 Excellent site

100×—The three brightest galaxies are just seen with averted vision. Using direct vision there is just a fuzzy spot at this location.

220×—With higher power and using the monk's hood, all five of the galaxies are held steady. NGC 7320 is the easiest to see; it is pretty faint, pretty small, much brighter in the middle and elongated 2×1 in PA 135. NGC 7317 is just south of a star and is just larger than the seeing disk of that star; averted vision makes it larger. NGC 7318A and B are only spilt as separate galaxies at higher powers. NGC 7318B is somewhat larger than its companion and a little elongated 1.5×1 in PA O. NGC 7319 is faint, small, brighter in the middle and somewhat elongated 1.8×1 in PA 120.

36 in. f/5 S=7 T=8 Excellent site

320×—NGC 7320 has very mottled arms and a knot in the arm to the north of a bright core. NGC 7318A and B are easily split and B has a pretty bright core. NGC 7317 is round and is about four times the size of the seeing disk. NGC 7319 really grows with averted vision; it also appears like a barred spiral, with a bright elongated section and much fainter outer portion. My best view of this group, ever (See Fig. 10.9).

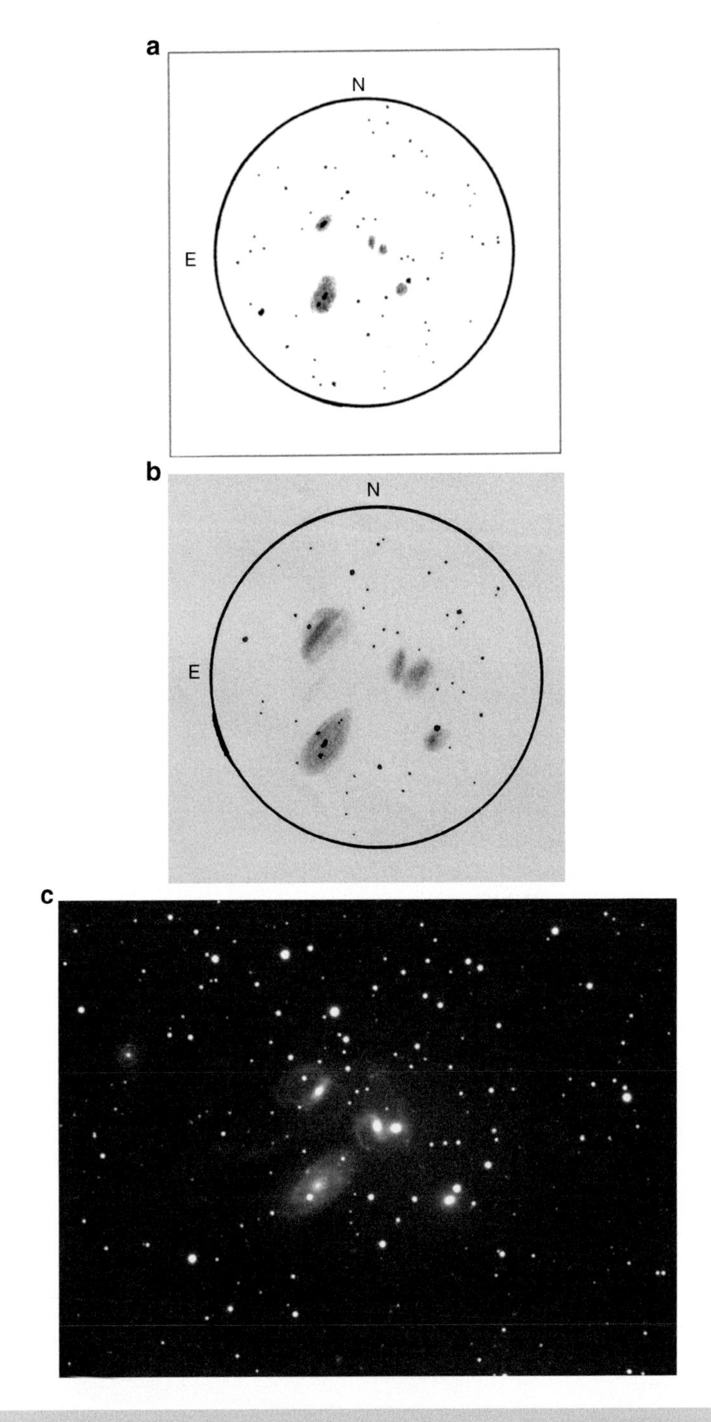

Fig. 10.9 (**a**) 13″ f/5.6; NGC 7320; FOV 20′; MAG 220×. (**b**) 36″ f/5; NGC 7320; FOV 15′; MAG 320×. (**c**) C 14 at f7-6 Atik 314L. *Photo: Parijat Singh*

Chapter 11

What Are All These Different Types of Nebulae, and What Details Can I See in Them with My Telescope?

There are five types of cloudy or nebulous objects in the sky: planetary nebulae, emission nebulae, reflection nebulae, dark nebulae and supernova remnants. I will cover the planetaries in the next chapter and discuss the other four types here. Even though all these objects appear as fuzzy and diffuse in the telescope, there are different mechanisms at work among the differing types of nebulae. So, let's see what makes each type of nebula glow in the dark (except dark nebulae, of course).

Emission Nebulae

The gases within these nebulae glow on their own; it is the actual gas cloud that is lit up to provide the photons that are gathered by your telescope. The method of exciting the atoms in a nebula to glow comes from the high-energy ultraviolet radiation being given off by hot stars within the nebula. This process is energetic enough to not just raise the energy levels of orbital electrons: it actually knocks them free from their parent atom. This is called ionization—the creation of a charged atom by removing one or more electrons. After a short period of time, the free electrons return to orbiting an atomic nucleus. This recombination releases radiation energy (a photon) as the electron returns to its place in orbit.

S.R. Coe, *Deep Sky Observing*, The Patrick Moore Practical Astronomy Series,
DOI 10.1007/978-3-319-22530-2_11

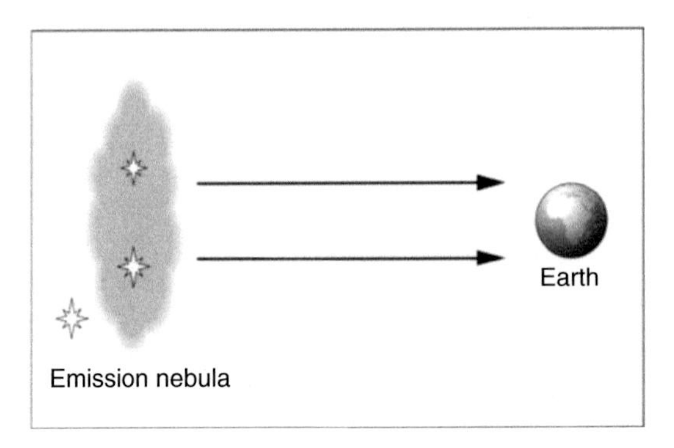

Fig. 11.1 An emission Nebula giving off light from the gas within the cloud

This light energy from recombination is at a specific wavelength, given the type of atom and amount of ionization that was produced by the incoming ultraviolet energy. So scientists can say with certainty that the two lines emitted in the green at about 5000 Å are from doubly ionized oxygen (O III) and nothing else. (An angstrom is a tenth of a nanometer, that is a tenth of a meter to the minus ninth; in other words an it is a meter to the minus tenth.)

Fig. 11.2 Eta Carina Nebula; 14″ SCT Hyperstar f/2. *Photo: Jim Barclay*

Singly ionized oxygen emits energy in the ultraviolet and doubly ionized nitrogen emits red light. "H alpha" radiation from hydrogen is a deep-red color, seen on many photographs of these emission nebulae (See Fig. 11.2).

Neon light tubes use this effect with electricity as a power source to light up advertising signs in Piccadilly Circus in London or Times Square in New York.

The reason that those nifty nebular filters work is that the coatings deposited onto the glass are very carefully controlled. Their chemical composition and thickness determine which narrow colors of light will pass and which will be blocked. When the coatings are done with precision, the filters block the light from streetlights and pass the light from nebulae.

Here are some data and observations of emission nebulae, some of them involved with clusters of stars.

Object	**NGC 281**
Type	**CL+NB**
Mag	**7.4**
Size	**4.0′**
Class	**n**
Constellation	**Cas**
RA	**00 52.8**
Dec	**+56 37**
Tirion	**1**
U2000	**36**
Description	**F, vL, dif, S triple * on np edge**

Because the human mind is so adept at visualizing shapes, observers will create names for many deep-sky objects from their outline. So, the North America nebula, the Swan nebula and this one, the "Pac-man nebula", are named for the shape of the nebulosity. If you get your eyes in the correct orientation, the famous electronic game's dot-eating creature will appear. I also see it as a thick comma.

6 in. f/6 S=7 T=8 Excellent site

25×—Very faint, pretty large, irregularly round, just a misty glow with four stars involved.

40×—Faint, pretty large; the star in the middle is seen as a double star. I can just see the "Pac-men" or fat comma shape. Averted vision brings out the nebula better.

40×+UHC—This is one of those nebulae that really respond to the nebular filter. The shape is now easy and obvious, the contrast is much better.

13 in. f/5.6 S=6 T=6 Good site

100×+UHC—Pretty bright, large, irregularly round, with 14 stars involved. It is just seen without the UHC. A dark lane intrudes into the nebula on the south side, forming the Pac-man shape.

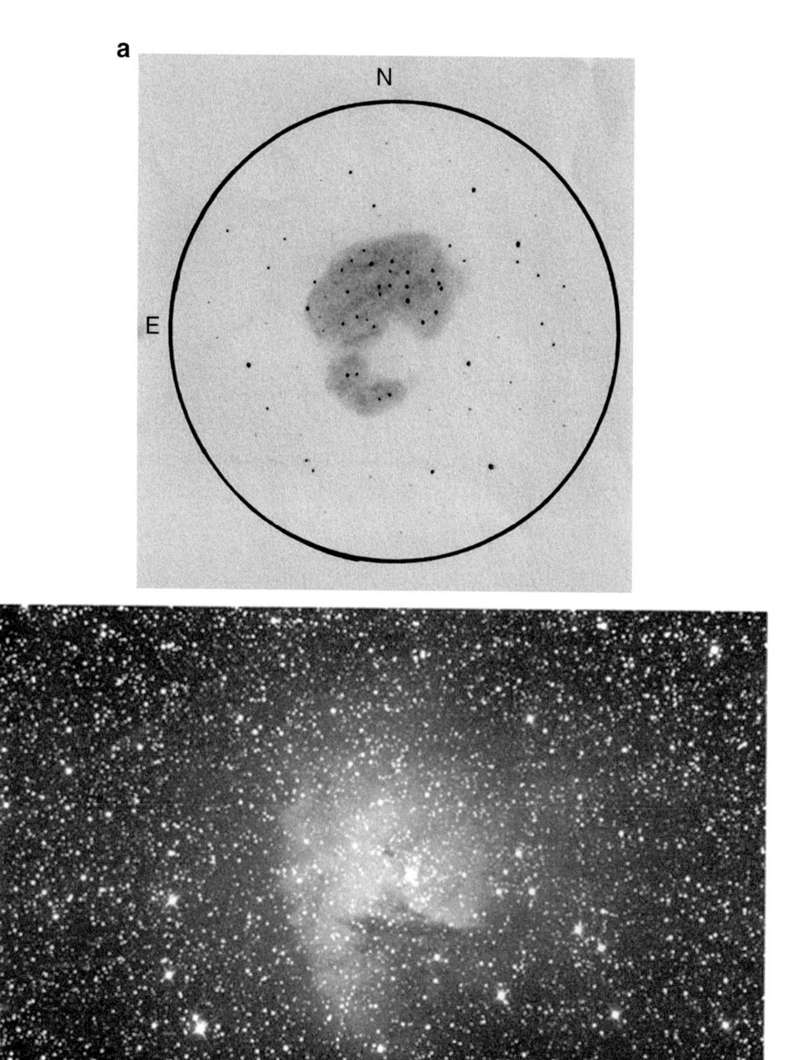

Fig. 11.3 (**a**) 13″ f/5.6; NGC 281; FOV 30′; MAG 100×. (**b**) 12″ f/5; 30 min exposure. *Photo: Chris Schur*

Object	**NGC 1491**
Other names	**H I 258**
Type	**BRTNB**
Size	**3′×3′**
Class	**E**
Constellation	**Per**
RA	**04 03.3**
Dec	**+51 18**
Tirion	**1**
U2000	**39**
Description	

Sometimes getting to a better site allows the UHC filter to work and increase the contrast and size of a nebula.

4 in.	f/8	S=6 T=8	Excellent site

40×—Pretty faint, pretty small, irregular figure, one star involved. Averted vision helps, but this nebula is still not much in small aperture.

13 in.	f/5.6	S=5 T=6	Mediocre site

100×—Pretty faint, pretty small, little brighter middle, 10th mag star on east edge, averted vision doubles the size, four 13th mag stars involved. 220×—Brings out two more stars for a total of seven. The UHC filter does not seem to increase the contrast or size of this nebula. Always gray in color

13 in.	f/5.6	S=8 T=9	Excellent site

135×+UHC—Pretty bright, pretty large, irregularly round. At 100× without the filter I just barely noticed this nebula. There is a 10th mag star involved and the UHC enhanced the contrast quite a bit.

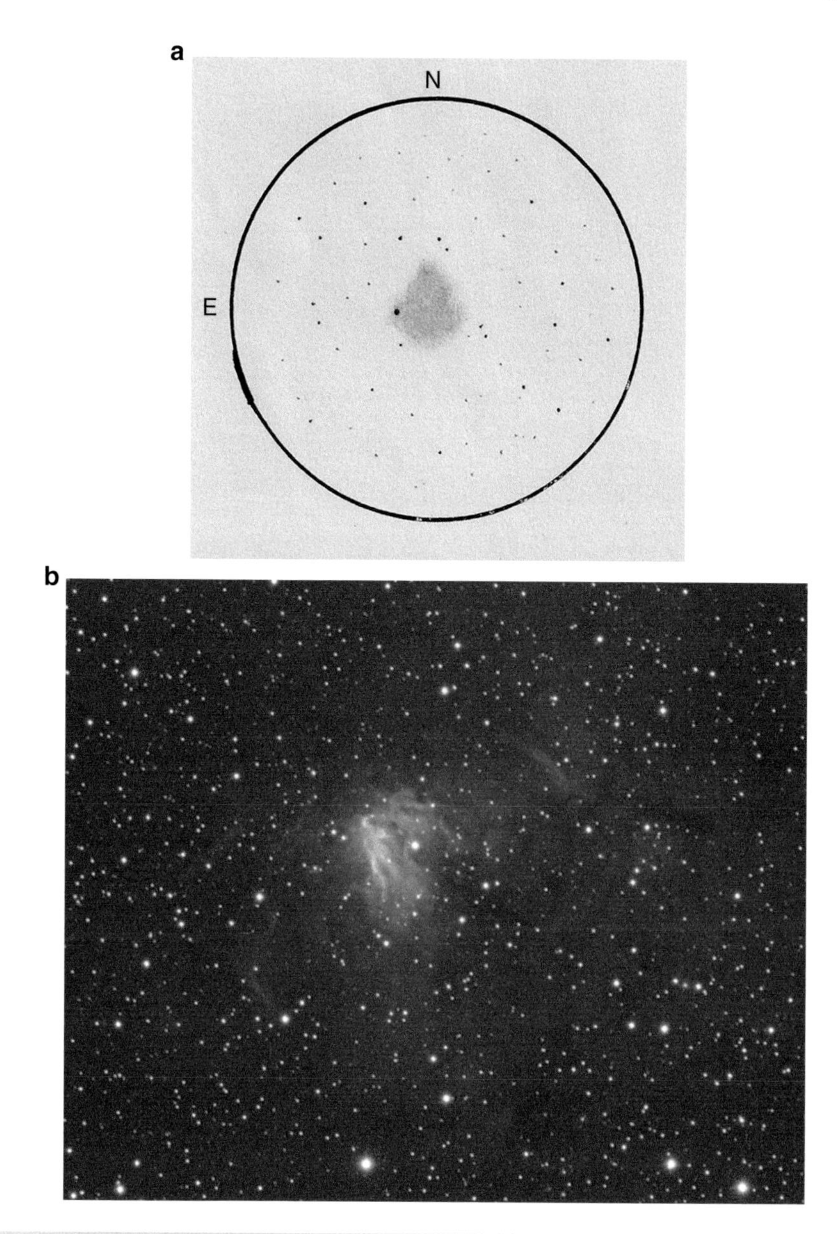

Fig. 11.4 (**a**) 13″ f/5.6; NGC 1491; FOV 20′; MAG 150×. (**b**) TEC 160 SBIG 8300 camera. *Photo: Lee Buck*

Object	NGC 1499
Other names	
Type	**BRTNB**
Size	**145′×40′**
Class	**E**
Constellation	**Per**
RA	**04 03.3**
Dec	**+36 25**
Tirion	**5**
U2000	**95**
Description	**vF, vL, mE ns, dif**

The California Nebula does indeed have the shape of that state on a good photograph, even on mine. However, that large shape is difficult to see with a telescope because of the narrow field of view. So, try a pair of binoculars or a finderscope from dark skies to see this large nebula.

11×80 finderscope S=7 T=8 Excellent site

11×—Pretty bright, very, very large, much elongated 2.5×1 in PA 75, with several stars involved. This wide-field view makes the "California" shape immediately obvious. Installing the UHC filter in the eyepiece of the 80 mm finder enhances the view considerably.

6 in. f/6 S=7 T=7 Excellent site

25×—Very faint, very large, much elongated. Adding the UHC filter raises the contrast very much. Now I can hold the nebula steady with direct vision. It is a much elongated glow that goes from one side of the field of view to the other, even with the giant Erfle eyepiece. I am observing by holding the UHC filter up to the eye lens of the giant eyepiece. The angle of the filter relative to the light exiting the eyepiece is quite critical. Very small amounts of tilt in the filter either darken the field of view considerably or do not provide the contrast enhancement of the nebula.

13 in. f/5.6 S=7 T=8 Excellent site

60×—Observing this object in the main scope is difficult because it is so large. Only one-third of the nebula will fit, even with a 38 mm eyepiece—which gives a 1° field. However, the increased aperture does show some light and dark bands lengthwise through the nebula.

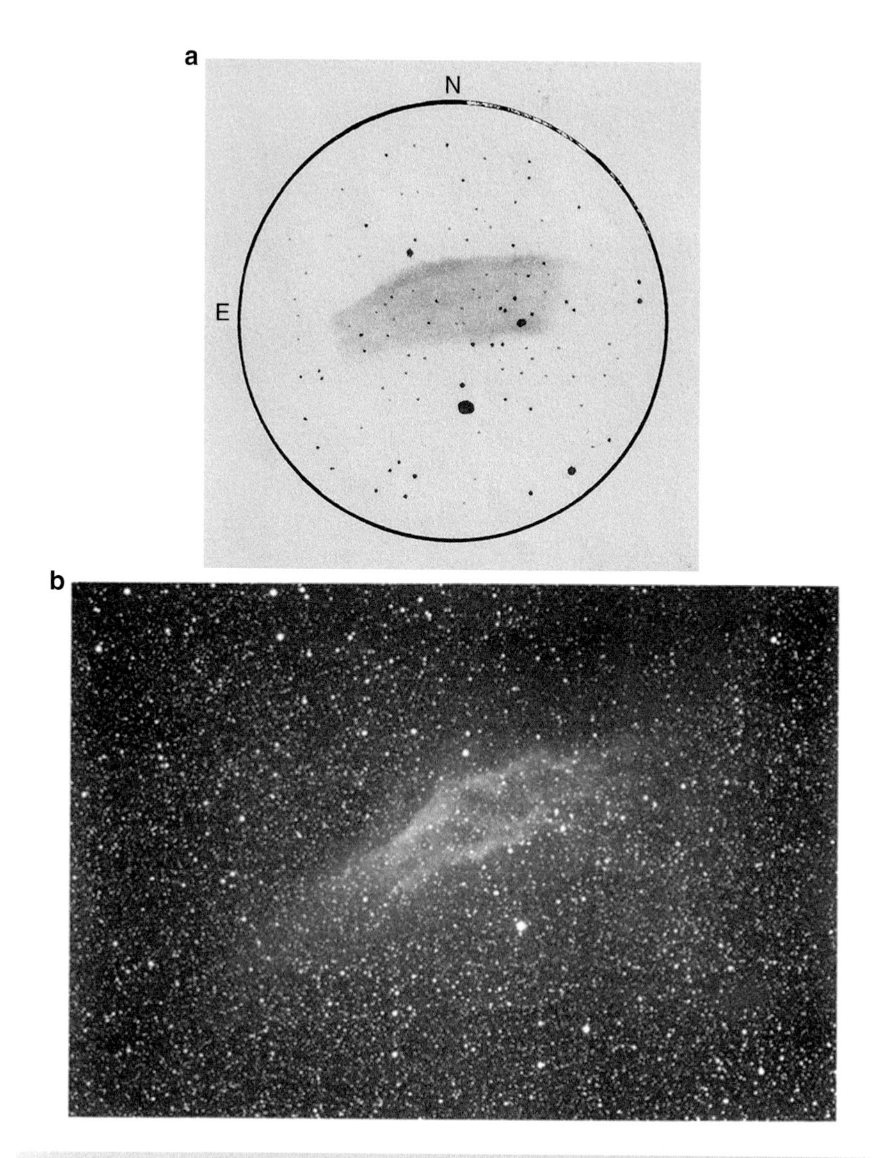

Fig. 11.5 (**a**) 11×80 finder, NGC 1499; FOV 3°; MAG 11×. (**b**) 8″ Schmidt cameral; 8 min exposure. *Photo: Chris Schur*

Object	**NGC 1999**
Other names	**H IV 33**
Type	**BRTNB**
Size	**16′ × 12′**
Class	**E + R**
Constellation	**Ori**
RA	**05 36.5**
Dec	**–06 43**
Tirion	**11**
U2000	**271**
Description	***10,11 inv in neb**

13 in.	f/5.6	S=7 T=7	Good site

220×—Pretty bright, pretty small, irregularly round, somewhat brighter in the middle, stellar nucleus, with several dark markings. This object is brightest on the north side and averted vision makes it grow in size.

25 in.	f/5	S=7 T=9	Excellent site

250×—The nebula is as large as the field of view, and has several striations of light blue within the gray nebulosity. There are very faint connections of nebulosity that connect M42–43 to NGC 1999; I have never seen those before.

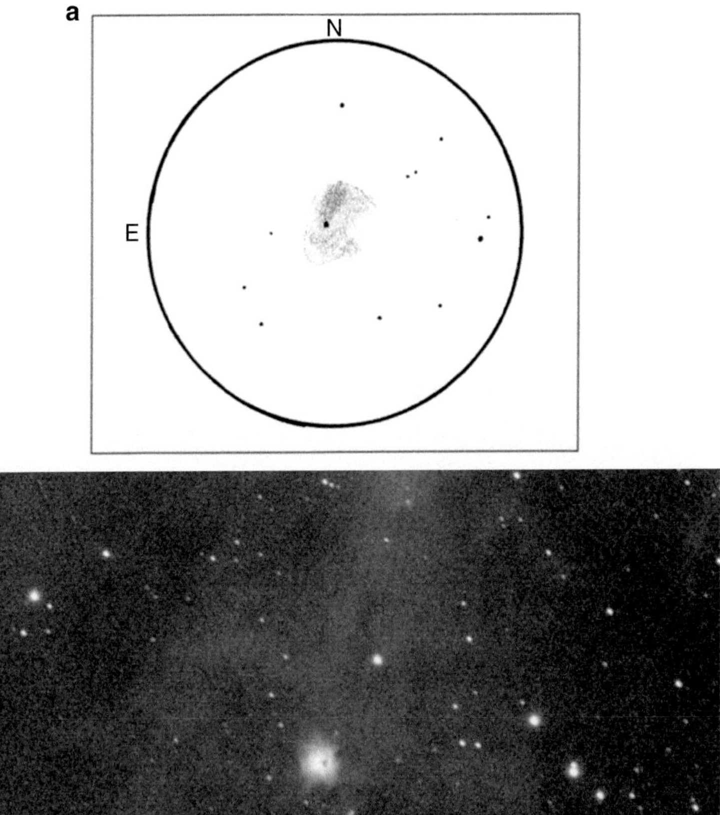

Fig. 11.6 (**a**) 13″ f/5.6; NGC 1999; FOV 20′; MAG 220×. (**b**) 8″ SCT at f/6. *Photo: David Douglass*

Object	**NGC 2261**
Other names	**H IV 2**
Type	**BRTNB**
Size	**2′×1′**
Class	**E(R)**
Constellation	**Mon**
RA	**06 39.2**
Dec	**+08 44**
Tirion	**11**
U2000	**182**
Description	**B, vmE 330°, Nucleus=*11**

This is Hubble's Variable Nebula. Edwin Hubble took many photographs of this fan–shaped nebula and stated that he could see changes in it over time. The star involved within the nebula is a variable star, R Mon; it is most likely that the changing light pattern from that star is affecting the emission output from the nebula. There are a set of photographs in *Burnham's* that show the changing aspect of NGC 2261. This object is the answer to a trivia question: it was the first deepsky object photographed with the 200 in. telescope at Mount Palomar.

4 in. f/8 S=6 T=8 Excellent site

80×—Faint, pretty small, very irregular figure, triangular or comet shaped. There is a faint star involved; it is more easily seen with direct vision.

13 in. f/5.6 S=7 T=9 Excellent site

135×—Bright, pretty large, much elongated and has a bright star involved. It appears as a small comet and the star R Mon is very obvious at the tip. The south side is brighter and the west side is more elongated.

330×—There is a dark lane on the south end; it cuts between the star R Mon and the rest of the nebulosity. The west side of the nebula is much brighter than the east side. Averted vision makes this nebula grow in size.

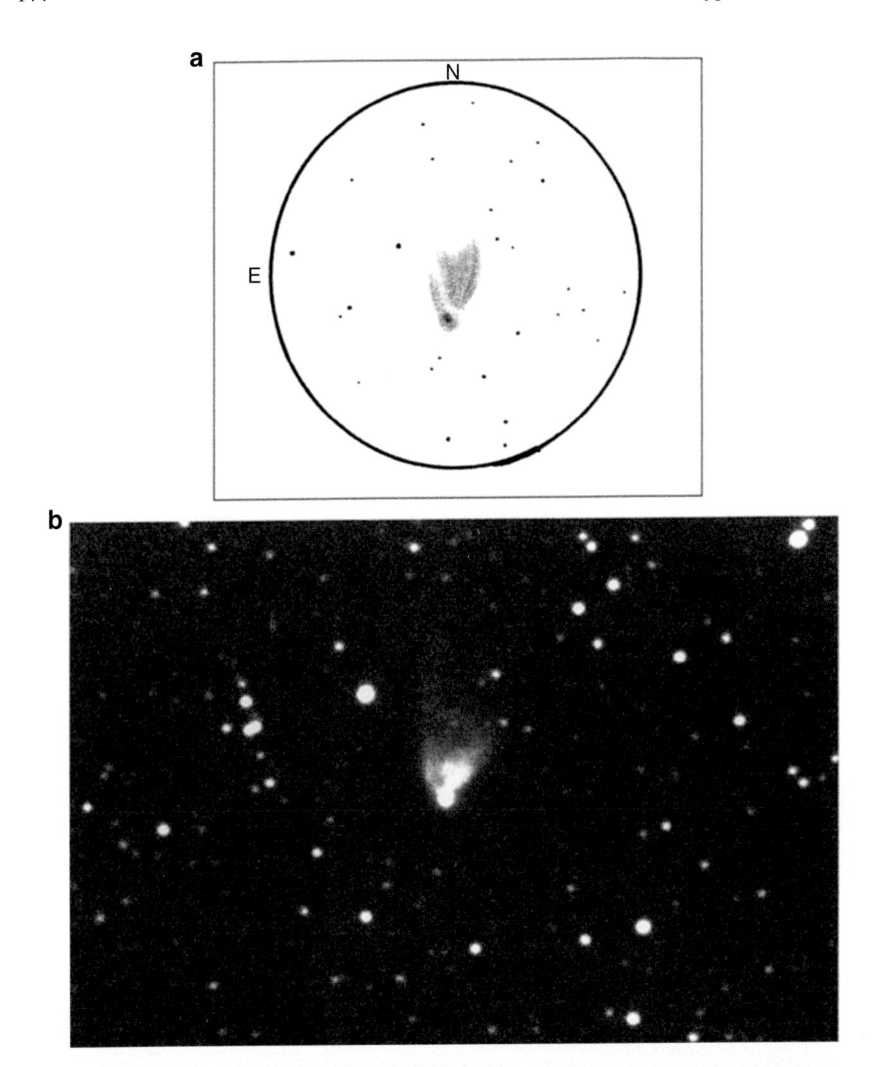

Fig. 11.7 (**a**) 13″ f/5.6; NGC 2261; FOV 10′; MAG 330×. (**b**) 12″ L×200 at f/7; ST 7E CCD; 6 min exposure. *Photo: Larry E. Robinson*

Object	**NGC 2359**
Other names	**H V 21**
Type	**BRTNB**
Size	**8.0′**
Constellation	**CMa**
RA	**07 17.8**
Dec	**–13 13**
Tirion	**12**
U2000	**274**
Description	**!!, vF, vvL, viF**

I have always heard NGC 2359 called the "Duck Nebula" because it has the shape of a duck's head seen from the side. The "eye" of the duck is a Wolf–Rayet star. These are hot, massive stars which have bizarre spectral characteristics. The emission line spectra are interpreted to mean that Wolf–Rayet stars are spewing huge amounts of material into interstellar space. These huge stars are driving a powerful solar wind that forms a nebula around them. The hot core that is left behind heats up the gaseous layers that form the nebula and they glow from ionization.

4 in. f/8 S=6 T=8 Excellent site

60×—Faint, pretty large, very irregular figure, the "Duck" is just seen. Adding the UHC filter shows the Duck better, with two stars involved. There is very little nebulosity beyond the Duck figure.

13 in. f/5.6 S=7 T=7 Excellent site

11×80—just barely seen.

100×—Pretty bright, large, irregular figure, "duck's head" shape seen immediately even without UHC filter.

100×+UHC—Adding UHC makes the field explode with nebulosity, many wisps covering it from edge to edge.

150×+UHC—Lots of detail with in the duck's head, several dark areas and stars involved; this is the best view; much nebulosity around the duck's head and detail within; there is a bright edge to the nebula on the north side.

36 in. f/5 S=7 T=9 Excellent site

165×+UHC—Shows as much detail as I have ever seen photographed in this object. 14 stars involved and a large nebulosity extends out of the field in several directions. Some parts of the nebula are almost spiral in shape. The detail around the eye of the duck, or central star, is amazing; there is mottling all around it. 250×—Shows a wealth of detail around the central part of this nebula; small bright areas and dark lanes swirl throughout this object with stars and bright knots also throughout. Easily my best view of this object (See Fig. 11.8).

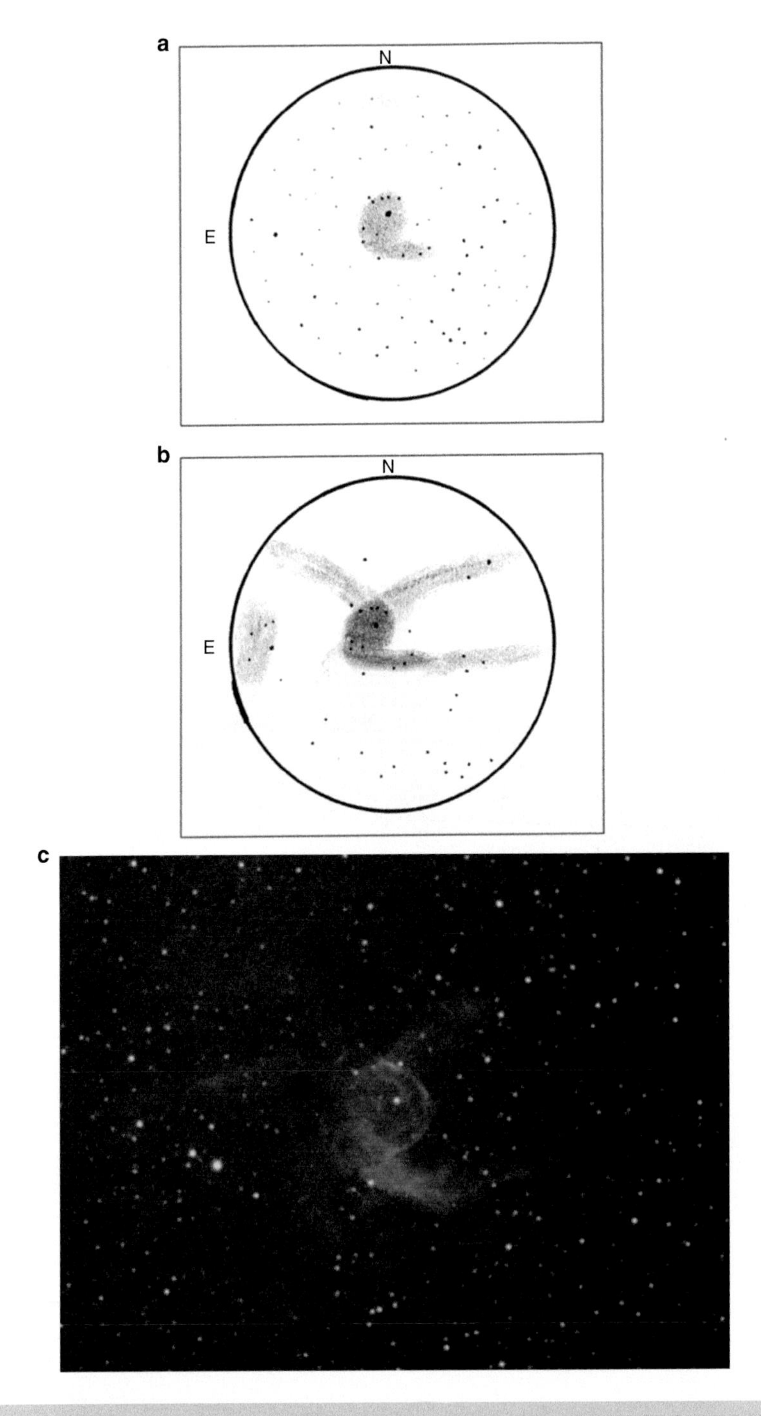

Fig. 11.8 (**a**) 13″ f/5.6; NGC 2359; FOV 20′; MAG 150× no filter. (**b**) 13″ f/5.6; NGC 2359; FOV 20′; MAG 150× with filter. (**c**) 8 in. SCT at f/6. *Photo: David Douglass*

Object	NGC 6334
Type	CL+NB
Size	20.0′
Number of stars	
Surface brightness	
Constellation	Sco
RA	17 20.6
Dec	–36 04
Tirion	22
U2000	376
Description	vB, S, iF, bM, r,* inv

6 in. f/6 S=7 T=8 Excellent site

40×—Very faint, large, irregularly round. There are 20 stars resolved in and around the nebula, most pretty faint, with a very faint background glow.

40×+UHC—Makes the nebulosity more obvious, but only pretty faint. There are several "clumps" or bright spots within the dim glow. There are also 12 stars involved within the nebulosity.

17.5 in. f/4.5 S=8 T=9 Superior site

100×—Faint, large, irregular figure. This emission nebula is the size of the 30′ field in a 20 mm Erfle eyepiece. The nebulosity has 15 stars involved, magnitudes 8–12. There are more stars involved on the eastern side of the nebula, so that appears to be brighter, but it is an effect of the stars involved, not brighter nebulosity.

100×+UHC—The filter brings up the contrast of this object. There are four bright areas of the nebula, with the entire field aglow with dim nebulosity.

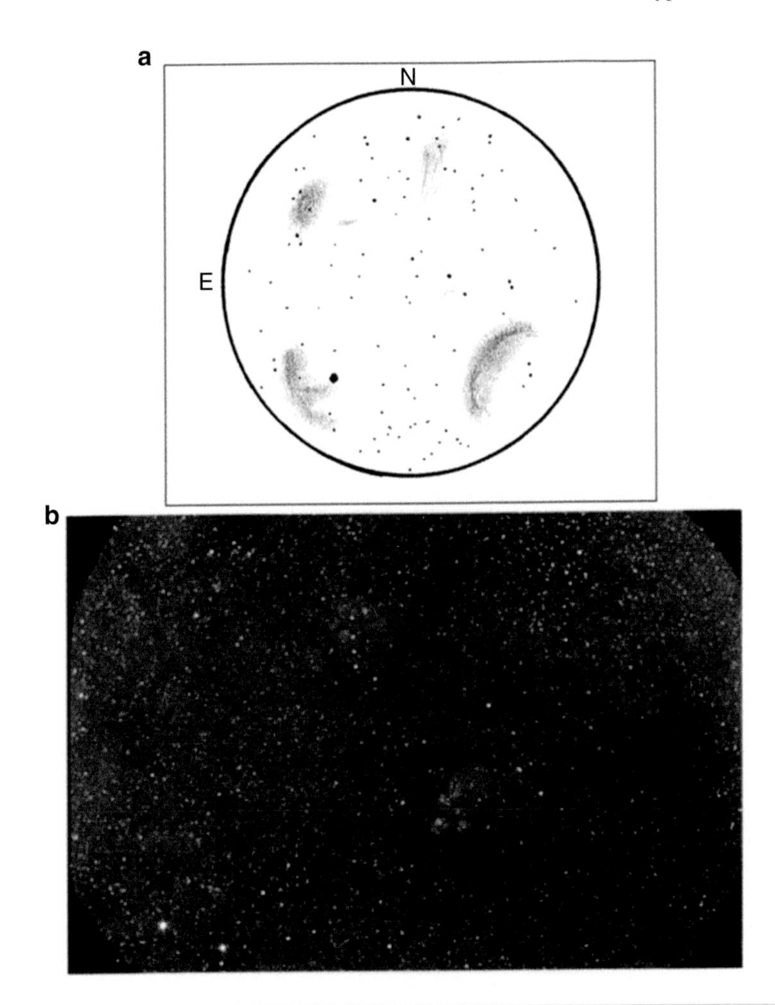

Fig. 11.9 (**a**) 13″ f/5.6; NGC 6334; FOV 30′; MAG 100× with UHC filter. (**b**) 8″ Schmidt camera; 7 min exposure. *Photo: Chris Schur*

Object	**NGC 6888**
Other names	**H IV 72**
Type	**BRTNB**
Size	**20′ × 10′**
Class	**E**
Number of stars	
Surface brightness	
Constellation	**Cyg**
RA	**20 12.8**
Dec	**+38 20**
Tirion	**9**
U2000	**119**
Description	**F, vL, vmE,** ** **att**

This is the Crescent Nebula, another name from an obvious shape. It is another nebula formed by a Wolf–Rayet star, in the center of the arc of nebulosity.

6 in. f/6 S=7 T=8 Excellent site

40×—Extremely faint, just a hint of nebulosity without the UHC filter; adding the filter raises the contrast considerably.

65×+UHC—Very faint, pretty large, irregular figure. An arc of nebulosity with four stars involved. The faintest section is to the southwest. Turning on the red light to draw it makes it disappear! The monk's hood helps the contrast somewhat. This is a very faint object with the UHC filter, the monk's hood and on an 8 out of 10 night. That is a dim object in 6 in. of aperture.

13 in. f/5.6 S=7 T=8 Excellent site

100×—Without UHC filter, just seen as a very faint nebula with some pretty bright stars involved. Adding the UHC makes a big difference; the nebulosity is now half of the field of view. It is pretty bright, pretty large, has an irregular figure—it is the "Crescent" nebula indeed. There are 18 stars involved, including a nice triple star on the south end. The northern part of the arch is much brighter than the south. It looks like the star in the middle of the arch has a thin finger of nebulosity attached, which makes the overall shape like the numeral 3.

150×+UHC—There are three bright knots within the nebula at this higher power. Also, I can see now that the central star is NOT attached to the Crescent of nebulosity. However, the illusion persists at lower powers.

36 in. f/5 S=6 T=8 Excellent site

200×+UHC—The arch of nebulosity takes up entire field of view. There is a knot of eight stars on south side. There are areas of smooth nebulosity and areas that are lumpy and irregular, a great view.

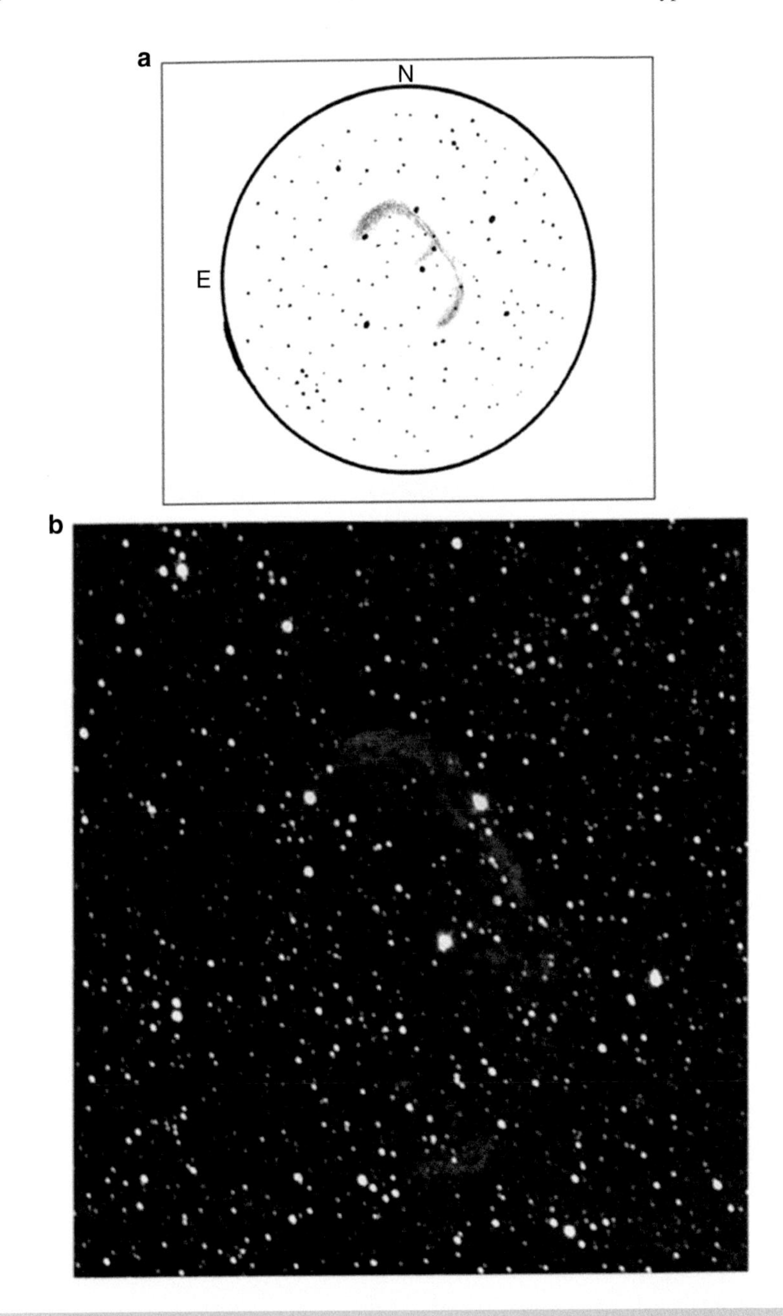

Fig. 11.10 (**a**) 13″ f/5.6; NGC 6888; FOV 30′; MAG 100× with UHC filter. (**b**) 20″ f/5; 20 min exposure. *Photo: Pierre Schwaar*

Object	NGC 7000
Other names	**H V 37?**
Type	**BRTNB**
Size	**120′ × 100′**
Class	**E**
Constellation	**Cyg**
RA	**20 58.8**
Dec	**+44 20**
Tirion	**9**
U2000	**85**
Description	**F, eeL, dif nebulosity**

This is the North America nebula, a large glow in Cygnus, near Deneb. The open star cluster NGC 6997 is involved within the nebula.

Object	NGC 6997
Other names	**H VIII 58**
Type	**OPNCL**
Constellation	**Cyg**
RA	**20 56.5**
Dec	**+44 39**
Tirion	**9**
U2000	**85**
Description	**Cl, P, IC, st L**
Notes:	**Scattered Cl in west part of North America Nebula.**

Naked eye S=7 T=8 Excellent site

1×—With no optical aid, I can just see a glow to the west of Deneb. It is difficult to know that it is the nebula—there are lots of stars in that region of the sky. It is probably the combination of the nebula and the stars that create this faintly glowing region.

10×50 binoculars S=7 T=8 Excellent site

10×—Faint, large, "North America" figure seen, better with averted vision. With the binoculars on a tripod for stability, I can easily see the shape of the nebula and resolve hundreds of stars in the field of view, they are most dense on the north side.

11×80 binoculars S=7 T=7 Excellent site

11×—North America shape is quite easy. I estimate that 50 stars are involved within the nebula and many hundreds of stars surround it. The cluster NGC 6997 is easy, but not resolved. However, what really stands out is the dark lane to the south of the famous shape. This dark area defines the southern outline of the North America shape and is striking with the big binoculars.

6 in. f/6 S=7 T=8 Excellent site

25×—Pretty faint, extremely large, irregular figure, many stars involved. The entire North America outline can be seen, with "Mexico" most obvious. The "Pelican" nebula is seen, but not nearly as bright as the North America.

40×—The cluster with in the nebula (NGC 6997) is pretty bright, pretty large, somewhat compressed, resolves into 18 stars, eight of which are very faint. At this power, the entire shape is larger than the field of view, but the brighter areas can be seen.

40×+UHC—The nebulosity does respond nicely to the UHC filter. Some light and dark bands within the nebula are seen in the brightest regions. Even with the filter in place, there are lots of bright and pretty bright stars involved within the nebulosity.

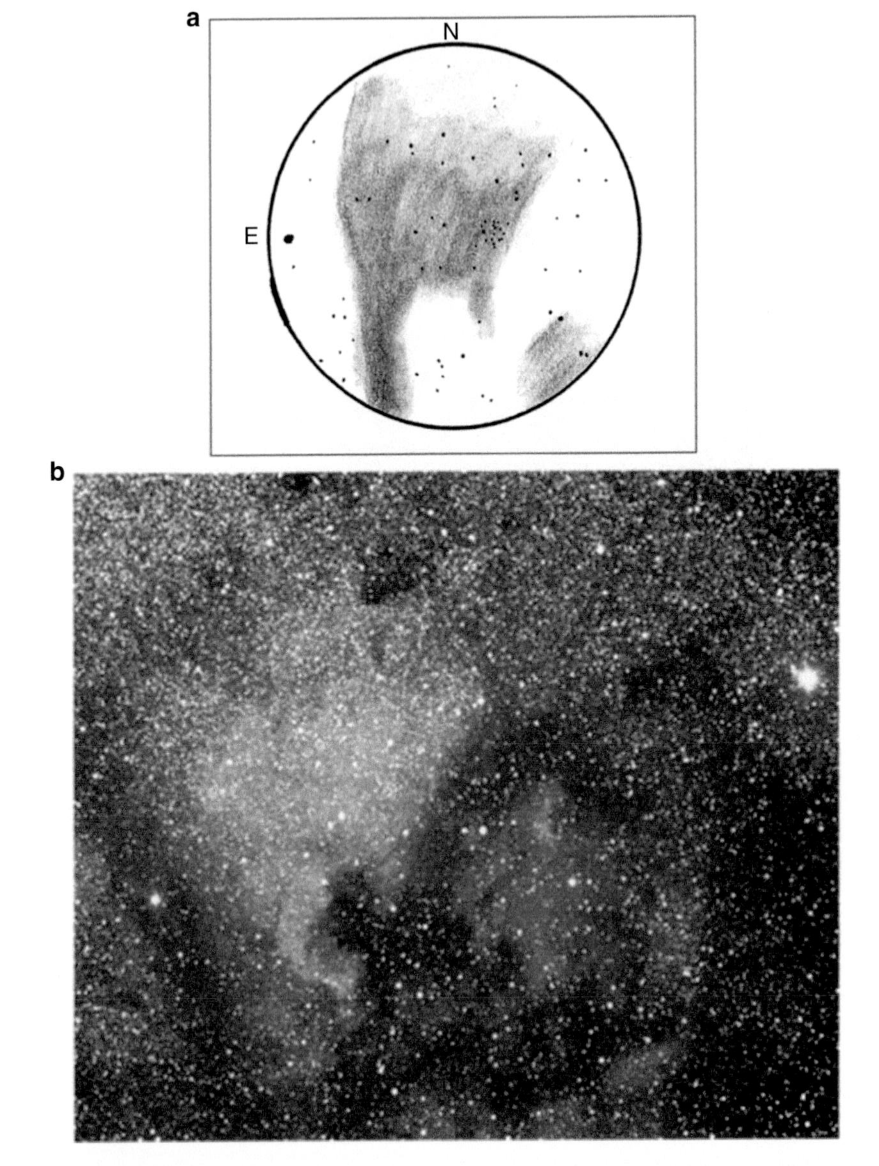

Fig. 11.11 (**a**) 6″ f/6; NGC 7000; FOV 1.5°; MAG 25× with UHC filter. (**b**) 8″ Schmidt camera. *Photo: Chris Schur*

Fig. 11.12 A wide-field photo of Cygnus, showing the North American nebula at the *top* (to the *left* of the bright star Deneb)

Object	**NGC 7380**
Other names	**H VIII 77**
Type	**CL + NB**
Mag	**7.2**
Size	**12.0′**
Class	**III 3 p n**
Number of stars	**40**
Brightest	**08.6**
Constellation	**Cep**
RA	**22 47.0**
Dec	**+58 06**
Tirion	**3**
U2000	**58**
Description	**Cl, pL, pRi, lC,*9…13**

6 in. f/6 S = 6 T = 6 Fair site

40×—Pretty bright, irregular triangular shape, not brighter in the middle. Averted vision "fills in" the triangular shape, 11 stars are counted. The cluster is very little compressed. 60 × 21 stars counted, 5 are at the limit of the 6 incher. The eastern side has a nice chain of stars of 11th magnitude. Adding the UHC filter shows a very little amount of faint nebulosity, even from a mediocre site.

13 in. f/5.6 S = 7 T = 8 Excellent site

100×—Pretty bright, pretty large, not compressed, triangularly shaped cluster of 36 stars. There is nebulosity involved in the cluster, which can be seen without the UHC filter, but the filter helps the contrast very much. The nebula is dim enough to disappear for a few moments when I return to the eyepiece after using my red flashlight to take notes (See Fig. 11.13).

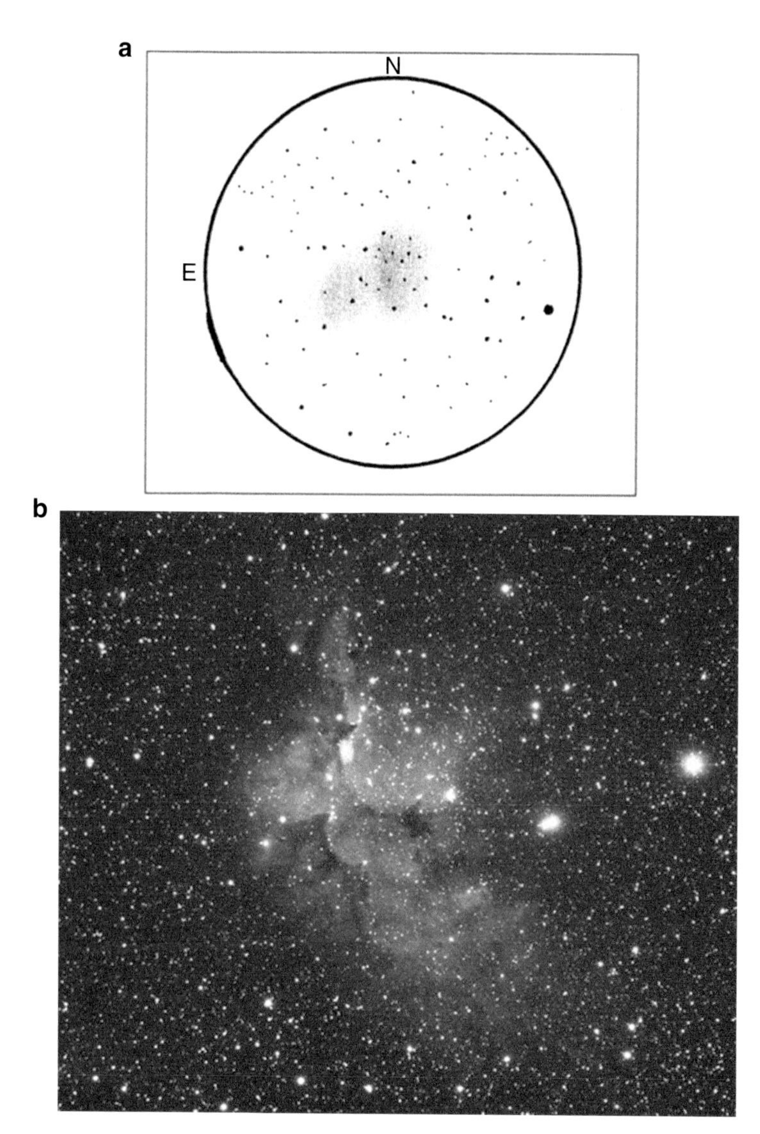

Fig. 11.13 (**a**) 13″ f/5.6; NGC 7380; MAG 100× with UHC filter. (**b**) TEC 140 f-7 QHY8 camera. *Photo: George Kolb*

Object	**NGC 7635**
Other names	**H IV 52**
Type	**BRTNB**
Size	**15′×8′**
Class	**E**
Constellation	**Cas**
RA	**23 20.7**
Dec	**+61 11**
Tirion	**3**
U2000	**34**
Description	**vF,* 8 inv, eccentric**

This is the Bubble Nebula; owing to the angle at which we are viewing the nebulosity, we see this "soap bubble" shape on long-exposure photographs—notice that a visual observation shows only an arc.

6 in. f/8 S=6 T=7 Good site

6×—Faint, pretty large, irregular figure with six stars involved in a low surface brightness glow. The UHC filter is no help; you need some photons to filter.

13 in. f/5.6 S=7 T=8 Excellent site

100×—Very faint, pretty large, much elongated, 5×1 in a PA of 90. A curved arc of nebulosity.

100×+UHC—Much better contrast with the night sky. Some very faint striations seen. This is a difficult object without the filter, but thicker and longer with the UHC filter in place. Six stars involved, one pretty bright, the rest pretty faint (See Fig. 11.14).

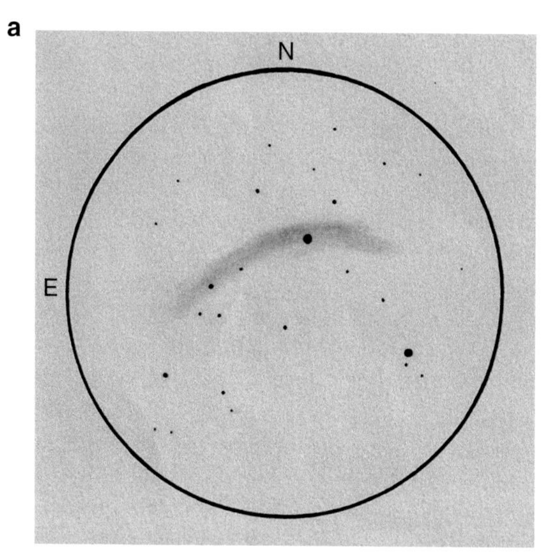

Fig. 11.14 (**a**) 13″ f/5.6; NGC 7635; FOV 30′; MAG 100×. (**b**) TEC 200 ED refractor ST-8300 camera. *Photo: Lee Buck*

Reflection Nebulae

These interstellar clouds are lit by the reflection of starlight. The tiny dust particles within the nebula reflect light from nearby stars and therefore the dust cloud glows. If you are sweeping the floor and there is sunlight streaming in from a window, then the dust is lit into a glow that can be easily seen. The exact same effect is happening to allow a reflection nebula to appear as a ghostly glow in the sky.

So, there must be a bright star embedded within or nearby which will illuminate these dust clouds. So, the starlight is bouncing off the dust grains in this interstellar cloud and then is seen on the Earth as a faint glowing nebula. Unlike an emission nebula, the glow is not from the molecules within the nebula (See Fig. 11.15).

The most famous example surrounds the Pleiades. This celebrated star cluster has reflection nebulae surrounding the brightest members. All long-exposure photographs of the Pleiades will display a glow around the bright stars in the picture. That the reflection nebula is a beautiful blue color is due to the scattering of the starlight as it reflects off the dust cloud. The longer red wavelengths are scattered away and reduced in intensity, leaving the blue wavelengths to provide the dominant color. The blue color is there in all reflection nebulae.

Several of the most famous and bright reflection nebulae also contain some emission portions as well. This combination means that the red emission and blue reflection sections of the nebula are easily distinguished in a color photograph. My example of a reflection nebula is of this combination type.

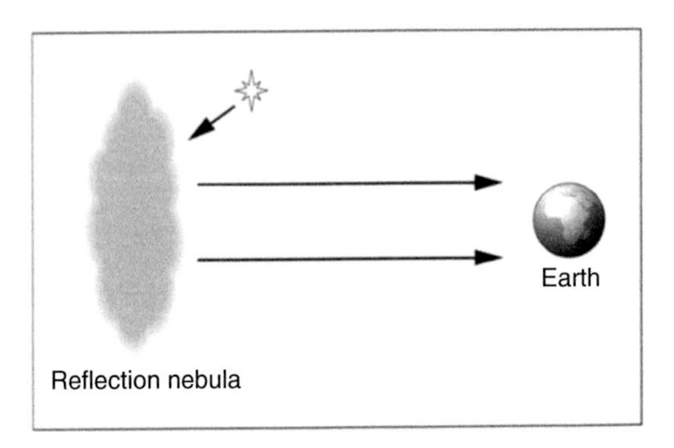

Fig. 11.15

Object	M20
Other names	NGC 6514
Type	CL + NB
Mag	6.3
Size	28.0′
Class	n
Number of stars	67
Brightest	06.0

(continued)

(continued)

Constellation	**Sgr**
RA	**18 02.3**
Dec	**–23 02**
Tirion	**22**
U2000	**339**
Description	**vB, vL, Trifid, D* inv**

Sir John Herschel named the Trifid Nebula for the dark lane that trisects the emission nebula. The reflection portion of this nebulosity is to the north. The combination of the red emission and blue reflection nebulae along with a prominent dark lane means that this object cannot be mistaken for any other.

11×80 binoculars S=7 T=7 Excellent site

11×—Both the emission and reflection portions of the nebula seen; only four stars resolved—averted vision makes it much larger. Overall, the Trifid Nebula is about one-tenth the size of the Lagoon Nebula.

6 in. f/6 S=7 T=8 Excellent site

65×—Bright, large, irregular figure. The Trifid shape is easy. There are 10 stars involved within the nebula and the dark lane structure is easy. The double star near the middle of the emission section (Herschel Number 40) is split. The pretty bright star inside the reflection portion of the nebula is light orange. The size of the reflection part of this nebula is doubled in size when using averted vision. A pretty prominent dark lane divides the two sections

65×+UHC—The reflection nebula is reduced to about half the size it was without the filter, but the emission section now shows much more contrast.

13 in. f/5.6 S=7 T=9 Excellent site

135×—Very bright, very large, irregularly round, approximately 40 stars involved in a nebula that is criss-crossed by dark lanes. The longest dark rift is to the north and it unequally divides the nebula, with a sector shape being formed by a dark lane to the west. The lanes meet at a faint, circular portion of the nebula. The reflection area to the north is quite fainter and round. There is an obvious and thick dark lane between the two regions. The UHC filter enhanced the nebulosity, but I like the view of this object better without the filter. The triple star in the center of the brightest section is HN 40; it is split at 165× and has two components which are yellow and a third which is light blue in color.

24 in. f/5 S=6 T=8 Excellent site

125×—Both portions of the nebula have lots of detail within them; I cannot see color, but the emission and reflection sections have a different "sheen" or texture, which was pretty easy to see once I noticed it. This is the most complex dark lane structure has ever been in any telescope. Wide, dark paths within the nebula have thin, winding fingers that stretch into the nebula. There are dark markings within the reflection part of the nebula, which are seen are elongated patches of darkness within the reflection glow.

With the 27 mm Panoptic and the 2 in. UHC filter, both parts of the nebulosity are easy to spot. I still cannot see color in them, but the smooth texture of the reflection nebula and the grainy or mottled appearance of the emission nebula are still easy to discern.

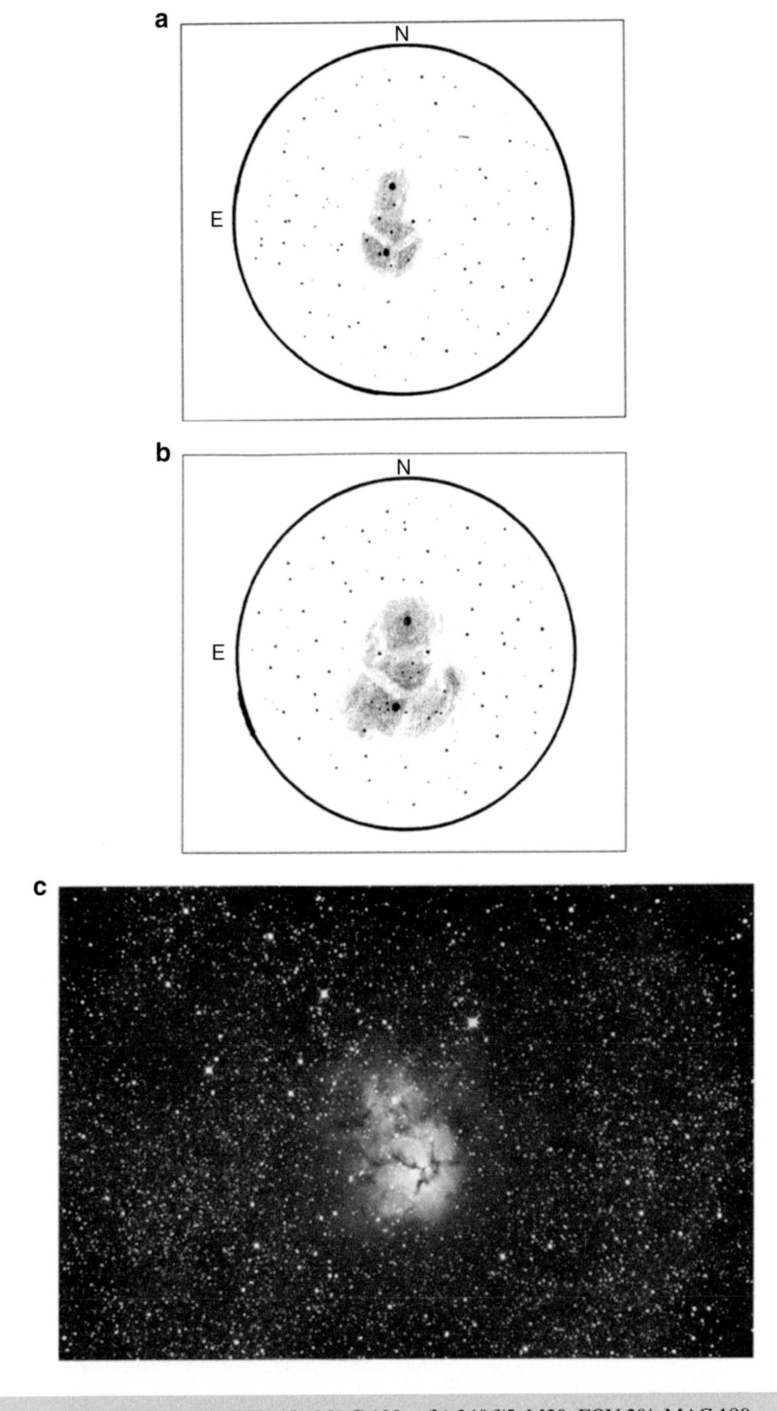

Fig. 11.16 (**a**) 6″ f/6; M20; FOV 45′; MAG 100×. (**b**) 24″ f/5; M20; FOV 20′; MAG 180× with UHC filter. (**c**) 12″ f/5; min. exposure. *Photo: Chris Schu*

Dark Nebulae (See Fig. 11.17)

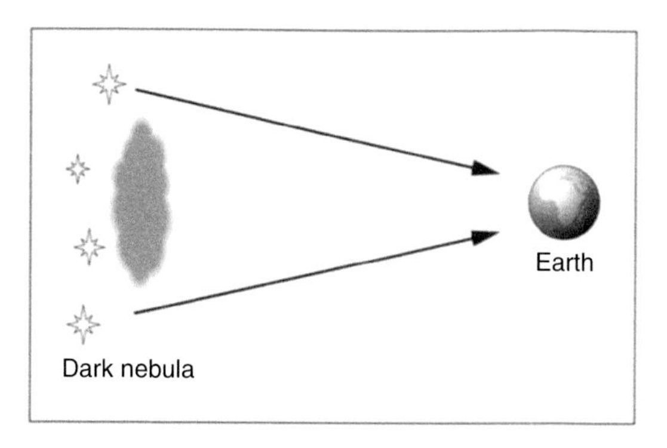

Fig. 11.17

It is certainly easy to define a dark nebula—it is a reflection nebula without the reflection. Both reflection and dark nebulae are clouds of stardust. Reflection nebulae have a star nearby to bounce light from the nebula to a terrestrial observer; dark nebulae do not have a star at the right location to make them shine, so they are dark. There are certainly lots of stars on the other side of a dark nebula; their light just can't get through the dark cloud (See Fig. 11.18).

Fig. 11.18 200 mm f/3.5 lens; 20 min exposure. *Photo: The Pipe Dark Nebula in Ophiuchus*

Don't take this to mean that these dark clouds are thick and soupy, quite the opposite—they are some of the most tenuous things that can be observed. No laboratory on Earth can reproduce the cold, thin and vacuous conditions that exist within a dark nebula. However, these clouds are large. Because it takes so long for a light beam from a star to get through these clouds, eventually the light is absorbed and the dark nebula remains visibly dark. Most of these interstellar clouds contain graphite, a form of carbon, just like in a pencil.

Observers know of these dark nebulae because they are silhouetted against the Milky Way. That bright background of Our Galaxy will show off any dark blocking clouds. E.E. Barnard photographed the swath of the Milky Way about a century ago and numbered many of the most obvious of these dark nebulae. So, all the dark nebulae I will be discussing have Barnard identification numbers assigned. Barnard also rated the opacity of the nebulae he discovered, giving numbers from 1 to 6. The higher numbers are the most opaque of the dark nebulae. Let's start with a very famous one, the Horsehead.

Object	**B33**
Type	**DRKNB**
Size	**6′×4′**
Class	**4**
Constellation	**Ori**
RA	**05 40.9**
Dec	**−02 28**
Tirion	**11**
U2000	**226**

Object	**IC 434**
Type	**BRTNB**
Size	**60′×10′**
Class	**E**
Constellation	**Ori**
RA	**05 41.0**
Dec	**−02 24**
Tirion	**11**
U2000	**226**
Description	**eF, vvL, vmE;1° long incl Zeta Ori**
Notes	**Contains Horesehead Nebula (B33)**

Object	**NGC 2023**
Other names	**H IV 24**
Type	**BRTNB**
Size	**10′×10′**
Class	**E(R)**
Constellation	**Ori**
RA	**05 41.6**
Dec	**−02 13**
Tirion	**11**
U2000	**226**
Description	**B* in Mid of L, lE neb**

Object	**NGC 2024**
Other names	**H V 28**
Type	**BRTNB**
Size	**30′ × 30′**
Class	**E**
Constellation	**Ori**
RA	**05 42.0**
Dec	**–01 50**
Tirion	**11**
U2000	**226**
Description	**! irr, B, vvL, black sp incl**

6 in. f/6 S=6 T=8 Excellent site

65×—This region is very "busy". There are lots of things to see, all within a field of view. The bright star Zeta Ori is in the field; to be able to see the most in each of these objects, move the scope so that Zeta is just outside the field of view; this makes a big difference. NGC 2024 is easy; even though it is just to the west of Zeta Ori, it is bright, large, irregularly round and has a thick dark lane cutting it almost in half. NGC 2033 is a faint, small, round nebula to the south. IC 434 is the emission nebula in the shape of a long, thin streamer that includes the Horsehead. With a small telescope, this is a difficult object. Get the star out of the field and try the UHC filter from a dark site. With all that done, I can see it as a very faint, pretty large, low-surface-brightness ribbon across the field of view. B33 is the Horsehead Nebula, a dark marking in front of IC 434. With the small scope I can only detect it with averted vision; it appears as a tiny notch missing from the nebula.

13 in. f/5.6 S=7 T=9 Excellent site

150×—Once the bright star is out of the field of view, then IC 434 is pretty faint and large. The Horsehead is a dark marking that cuts the streamer almost in half.

150×+UHC—The emission nebula is much easier with the filter in place. Therefore, B33 stands out better as a dark, rounded notch that protrudes into the nebulosity from the west. NGC 2024 is easy and the wide dark lane that divides it has light and dark markings that remind me of track marks left by a caterpillar tractor. Therefore, Arizona observers have taken to calling this object the "Tank Track" nebula. See if you agree.

36 in. f/5 S=6 T=8 Excellent site

200×—Wow, does aperture ever make a difference for the Horsehead! IC 434 is easy, bright and extends out of the field on both sides. The Horsehead shape is easy and shows detail, there is a "beard" of dark lanes that extend out from the mouth into the surrounding nebulosity. A lovely, delicate double star is at the neck of the horse. I am struck by the star density difference from west to east. The western side of the field has a myriad of faint stars that form nice chains and pairs. The eastern side of this field shows only five stars that are pretty faint and a black background.

200×+UHC—The dark Horsehead stands out better, but I liked the view without the filter; what makes a difference is the big telescope, not the little filter.

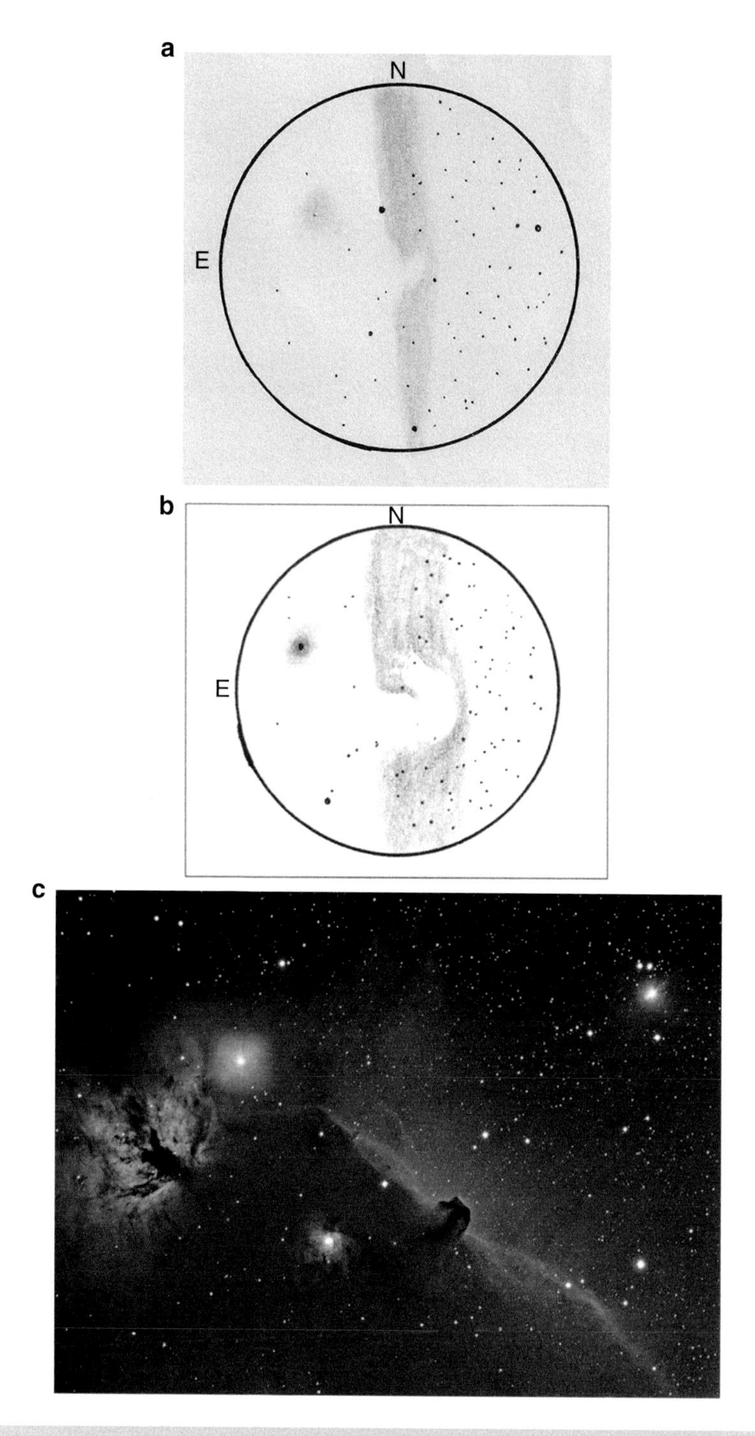

Fig. 11.19 (**a**) 13″ f/5.6 B33; FOV 25′; MAG 150× with UHC filter. (**b**) 36″ f/5; B33; FOV 12′; MAG 300×. (**c**) TEC 140 f-7 QHY8 camera. *Photo: George Kolb*

Object	**B86**
Other names	**LDN 93**
Type	**DRKNB**
Size	**4′**
Class	**5**
Constellation	**Sgr**
RA	**18 02.7**
Dec	**−27 50**
Tirion	**22**
U2000	**339**
Description	**Very distinct Drkneb**

Object	**NGC 6520**
Other names	**H VII 7**
Type	**OPNCL**
Mag	**7.6**
Size	**6.0′**
Class	**I 2 m n**
Number of stars	**60**
Brightest	**09.0**
Constellation	**Sgr**
RA	**18 03.4**
Dec	**−27 54**
Tirion	**22**
U2000	**339**
Description	**Cl, pS, Rl, IC, st9…13**

Of all the areas of the Milky Way I have observed, this region is unique. Once you have seen B86 and NGC 6520, a distinct dark place (the Ink Spot) and a nice open cluster in the same medium-power field of view, you will never mistake this for any other location in the sky.

6 in. f/6 S=7 T=8 Excellent site

25×—Just seen as a fuzzy area.

65×—This is a great view. The cluster resolves into seven stars that are held steady and another five stars that are seen with averted vision. The cluster is faint, small, very compressed and not rich. The dark spot is easy and obvious, it contains no stars. There is a nice orange star seen on the opposite side of B86 from the cluster. All in all, a terrific field of view.

13 in. f/5.6 S=7 T=9 Superior site

150×—Pretty bright, pretty large, pretty rich, compressed, 22 members resolved. There is a nice light-orange star in the center of the cluster. Dark nebula B86 is unmistakable to the west, just at the edge of the star cluster, and the dark nebula and the cluster are about the same size. There is also a pretty bright orange star at the edge of the dark nebula, across from the cluster. A fascinating field of view. There are only four very faint stars involved in the dark nebula. There is an elongated dark area south of the cluster.

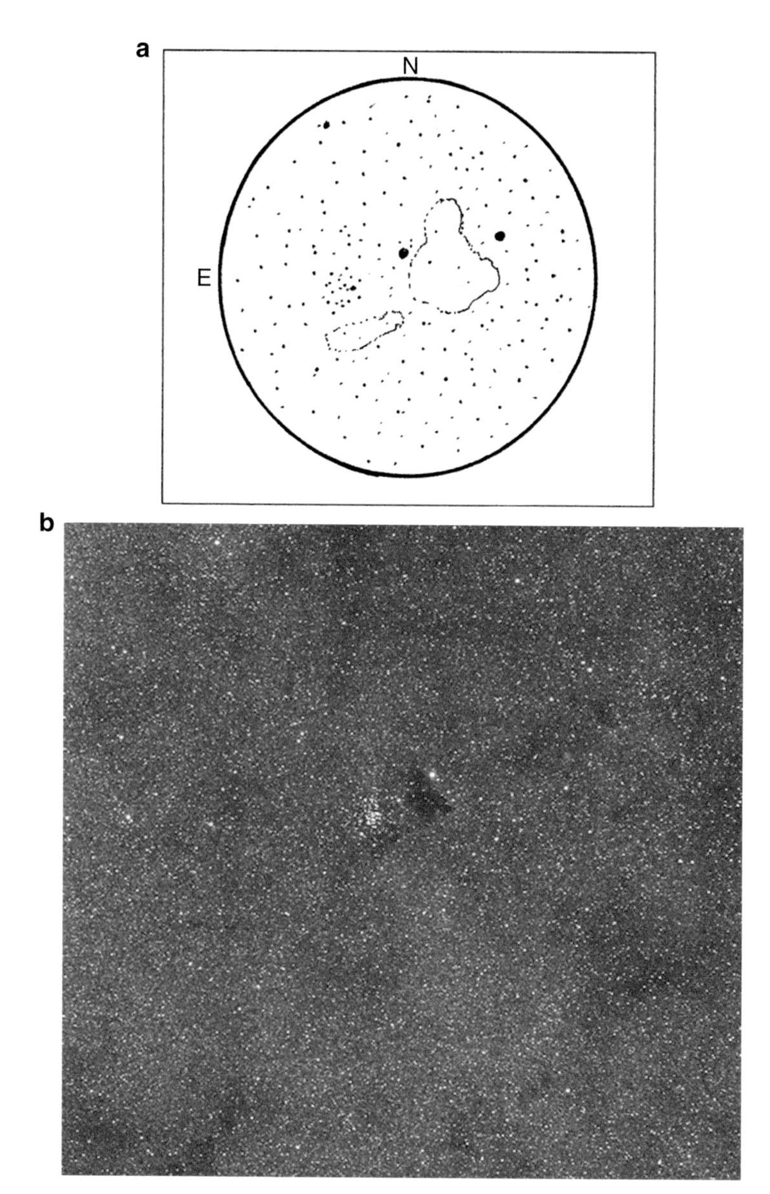

Fig. 11.20 (**a**) 13″ f/5.6; B86; FOV 25′; MAG 150×. (**b**) ED 80 f/6; 4 min exposure; Canon T2i camera

Object	B92
Other names	LDN 323
Type	DRKNB
Size	12′×6′
Class	6
Constellation	Sgr
RA	18 15.5
Dec	–18 11
Tirion	15
U2000	339

Object	B93
Other names	LDN 327
Type	DRKNB
Mag	N/A
Size	12′×2′
Class	4
Constellation	Sgr
RA	18 16.9
Dec	–18 04
Tirion	15
U2000	339

These two dark nebulae are within 30 arcmin of one another, and because B92 has a darkness rating of 6 and B93 has a rating of 4, you can easily see the difference in these two dark areas. They are on the northwest side of M24, the Small Sagittarius Star Cloud. I will cover M24 as a separate object in Chap. 13.

Naked eye S=6 T=8 Excellent site
1×—Even at one of my best sites on a good night, I cannot see these dark lanes without optical aid.

8×25 binoculars S=6 T=8 Excellent site
8×—The two round dark “eyes” on the top side of M24 are just seen. If I use averted vision with the small binocular, the dark markings can be held steady.

10×50 binoculars S=6 T=8 Excellent site
10×—The round dark areas on the north side of M24 are easy and obvious. They can be held steady with direct vision.

6 in.	f/6	S=6 T=8	Excellent site

40×—A fine view of B92 and 93. B92 is to the west and is the larger and more opaque of the two. It is round, large and only has one star within its boundaries. There are several other dark lanes that meander off to the west from this dark nebula. B 93 is noticeably smaller and less opaque, it is elongated N–S about 2× and there are a dozen stars that invade its borders.

Fig. 11.21 B 92 is the prominent dark oval on the north (*top*) side of M 24. B 93 is the elongated dark area to its *left*. Photo: ED 80 at f/6; 4 min exposure

Object	**B133**
Other names	**LDN 531**
Type	**DRKNB**
Size	**10′×3′**
Class	**6**
Constellation	**Aql**
RA	**19 06.1**
Dec	**−06 50**
Tirion	**16**
U2000	**296**

6 in. f/6 S=7 T=8 Excellent site
65×—There are no stars seen within an elongated dark spot. There are several other small dark areas nearby, but this is easily the most obvious.

13 in. f/5.6 S=7 T=8 Excellent site
100×—It is half the field in size and elongated 2×1 in PA 0°. There are four very faint stars involved within the dark area, but otherwise there are no stars seen. While you are looking in this area of the sky, the red star V Aql is nearby; at 100× it was a lovely dark-orange star afloat in a rich Milky Way field of view.

Object	**B134**
Other names	**LDN 543**
Type	**DRKNB**
Size	**6′**
Class	**6**
Constellation	**Aql**
RA	**19 06.9**
Dec	**–06 14**
Tirion	**16**
U2000	**296**
Notes	**1.4° south of Lamba Aql.**

6 in. f/6 S=7 T=8 Excellent site
40×—It can just been seen at low power as an area of missing stars
65×—Much easier; a round dark spot with some dark lanes to the NNW.
There are very few faint stars within the nebula. It is one of those objects that, once you see it, is really an easy one.

13 in. f/5 S=7 T=8 Excellent site
100×—An easy dark spot; no stars seen within a round area about 5 arcmin across. It is surrounded by stars of magnitudes 8–12.
150×—One very faint double star involved, magnitudes 12 and 13.

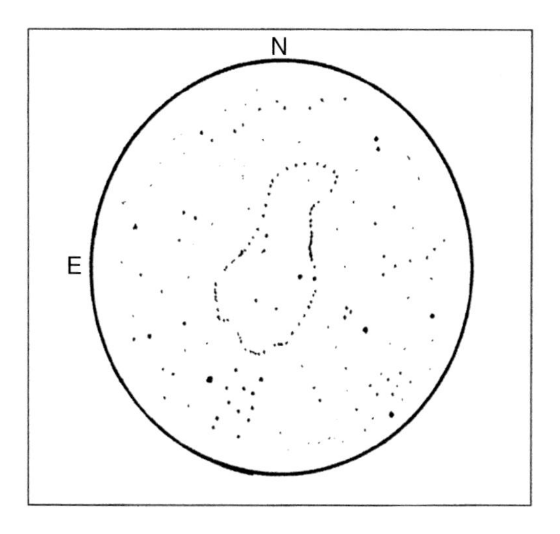

Fig. 11.22 13″ f/5.6; B133; FOV 30′; MAG 100×

Supernova Remnants

Stars of "normal" size and mass live out their lives with a steady, dependable output of light and heat. The Sun is such a star and it is the reason we have had the time to evolve under its generally constant supply of energy. At the end of their lives these stars pass through the planetary nebula stage and then very slowly cool.

However, the really big stars just can't go quietly. They use their fuel relatively quickly and that heat and energy creates a huge atmosphere around a core that is running out of elements to fuse together. The massive atmosphere of these giant stars eventually crashes onto the dense, hot core and an explosion of astronomical scale takes place. The light from that explosion is called a supernova and in some cases that one star equals the light of its parent galaxy for about a week. A prodigious amount of energy, even on the scale of the Cosmos.

The core of the progenitor star was a fusion engine before the explosion and it created a wide variety of different chemical elements. As one fuel was used up, it was fused into a new element, on up the periodic table. One of the most amazing things to know about how life is created is that the calcium in your bones, the iron in your blood, all the heavy elements in your body, were created within a heavy star and then slung out into the Universe by an ancient supernova explosion. More than any other fact, this amazing method of creating and distributing the raw materials of life compels me to see the Universe as the work of a Grand Plan.

Once the gigantic explosion has happened, what used to be the outer layers of the star are now propelled into space and these molecules come into contact with the interstellar material already there. Eventually, the gas and dust of the supernova remnant will dissipate and cease to be a glowing nebula. Like other nebulae, the gas glows from photons released from ionization and then recombination. The energy to strip electrons from the atoms in the supernova remnant comes from either the very hot core of the star, left behind after the explosion, or from nearby hot stars that are emitting lots of ultraviolet radiation.

So far the type of supernova that I have been explaining is called a "Type II supernova". This involves the large, massive stars which collapse and explode. There is another type of supernova that is brighter still, it is called a "Type I supernova" and the progenitor stars are white dwarfs that are part of a binary star system. The material from the "normal" star slowly falls onto, or accretes, over the surface of the white dwarf. This gain in mass means that eventually the white dwarf cannot maintain its size. The white dwarf star collapses in a few seconds and a gargantuan amount of energy is released as the supernova explodes. This massive blast blows the normal star apart and the dust and gas then form the nebula we see today.

Even though there are hundreds of cataloged supernova remnants, there are few available to amateur observers. Because supernovae are somewhat rare, the remnant material is usually discovered by long-exposure photographs.

However, several of the brightest are very famous deep-sky goodies and deserve the many hours spent studying them at the eyepiece.

Object	**IC 443**
Type	**SNREM**
Size	**60′×8′**
Class	**E**
Constellation	**Gem**
RA	**06 16.9**
Dec	**+22 47**
Tirion	**5**
U2000	**137**
Description	**F, narrow curved**

6 in. f/6 S=6 T=8 Excellent site

40×+UHC—Even with the monk's hood over my head to block extraneous light and after a short break from observing to get well dark-adapted, nothing was seen at the position of this object.

13 in. f/5.6 S=7 T=9 Excellent site

100×+UHC—Extremely faint, large, elongated, curved arc shape.

Seen only with UHC filter, a dim wisp. I would not have found this supernova remnant if it had not been for *Uranometria 2000* providing the exact position. Using the monk's hood to block out some stray light did help with the view somewhat, but this is still a difficult object.

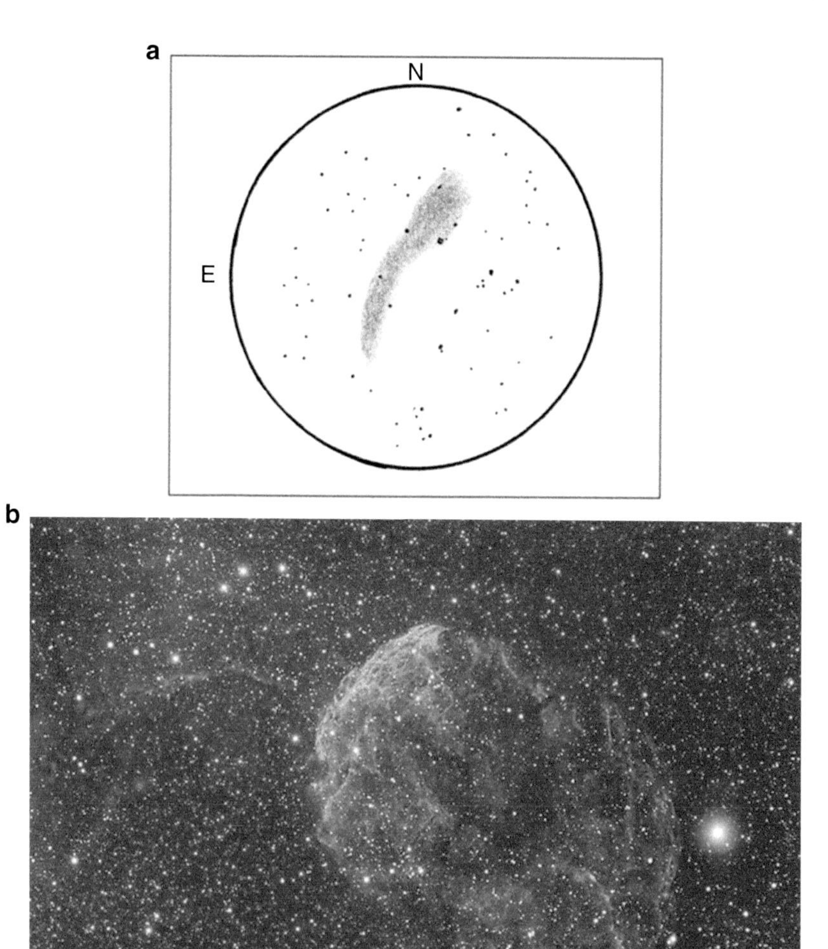

Fig. 11.23 (**a**) 13″ f/5.6; IC 443; FOV 30′; MAG 100×. (**b**) 160 Stellarvue APO refractor; *Photo: Mike Miller*

Object	**NGC 6960**
Other names	**H V 15**
Type	**SNREM**
Size	**70′ × 6′**
Class	**E**
Constellation	**Cyg**
RA	**20 45.6**
Dec	**+30 43**
Tirion	**9**
U2000	**120**
Description	**pB, cL, eiF, Kappa Cyg invl**
Notes	**Veil Nebula western part,* 52 Cyg invl.**

Object	**NGC 6974**
Type	**BRTNB**
Size	**60′ × 8′**
Constellation	**Cyg**
RA	**20 50.8**
Dec	**+31 52**
Tirion	**9**
U2000	**120**
Description	**neb*, neby cE pf**
Notes	**Portion of Veil Neb between two arcs**

Object	**NGC 6979**
Other names	**H II 206**
Type	**BRTNB**
Size	**60′ × 8′**
Constellation	**Cyg**
RA	**20 51.0**
Dec	**+32 09**
Tirion	**9**
U2000	**120**
Description	**vF, S, IE, sev F*f nr**

Object	**IC 1340**
Type	**SNREM**
Size	**60′ × 8′**
Class	**E**
Constellation	**Cyg**
RA	**20 56.2**
Dec	**+31 04**
Tirion	**9**
U2000	**120**
Description	**Possibly conn w NGC 6995**
Notes	**Veil nebula (See Fig 11.24).**

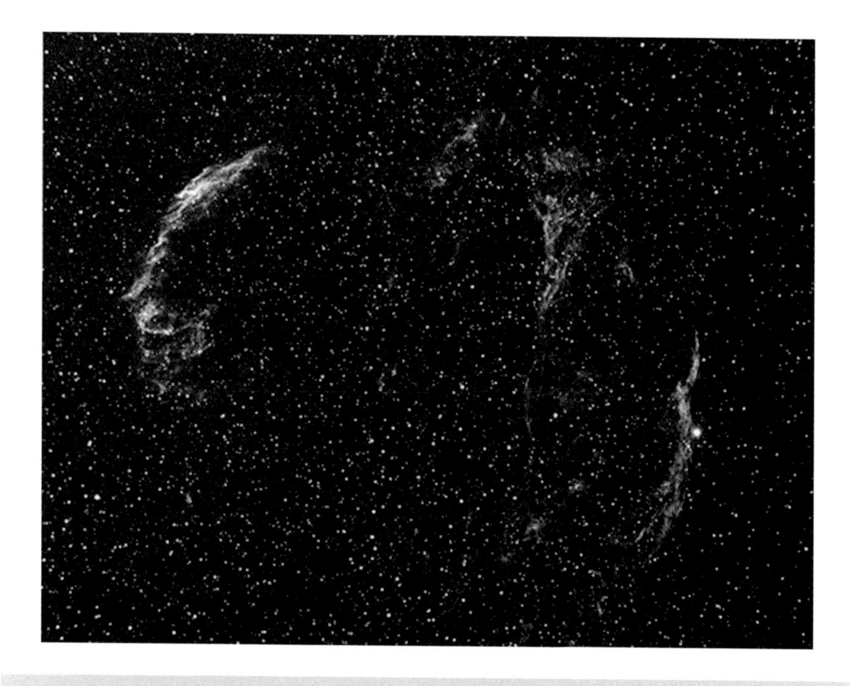

Fig. 11.24 8″ Celestron SCT, this is four frames assembled. *Photo: David Douglass*

Object	**NGC 6992**
Other names	**H V 14**
Type	**SNREM**
Size	**60′ × 8′**
Class	**E**
Constellation	**Cyg**
RA	**20 56.3**
Dec	**+31 42**
Tirion	**9**
U2000	**120**
Description	**eF, eL, eE, eiF, Bifid**
Notes	**Veil Nebula eastern part**

Object	**NGC 6995**
Type	**SNREM**
Size	**60′ × 8′**
Class	**E**
Constellation	**Cyg**
RA	**20 57.0**
Dec	**+31 13**
Tirion	**9**
U2000	**120**
Description	**F, eL, neb&st in groups**

NGC 6960 and 6992 are the brighter parts of the Veil Nebula. This entire nebula was created by a supernova about 30,000 years ago and we just happen to be lucky enough to live while it is visible. NGC 6960 passes behind 52 Cygni, a naked-eye star off the western wing of the Swan. NGC 6992 is about 2° from 52 Cygni and is somewhat brighter than 6960. Between these two sections is IC 1340, a triangular part of the nebula that was discovered by E.C. Pickering. I have heard it called Pickering's Wedge. This is one of those places in the sky where you do not want to hurry. There is lots to see; take your time and drink fully from the glass.

10×50 binoculars S=6 T=8 Excellent site

10×—Both 6960 and 6992 can be seen even at this low power; the star overwhelms 6960 somewhat, but the fact that you are seeing the broken pieces of what used to be a complete bubble is obvious. I call this the "Loop" effect and it is usually only seen in wide-field photographs. Here is yet another place where the binoculars create a view that cannot be duplicated in a telescope.

6 in. f/6 S=6 T=8 Excellent site

40×—NGC 6960 is easy to see, north of 52 Cygni. It appears as a pointed section of nebulosity that is quite bright. The section that forks to the south of the star is much more difficult and needs averted vision. Adding the UHC filter makes a dramatic difference; the fork section is now easily held with direct vision. The same is true of NGC 6992: I can see it with the RFT and no filter, but the addition of the UHC makes a dramatic difference in the amount of detail that can be seen. Many stars of magnitudes 9–12 are involved within the nebula. Either 6960 or 6992 will fit in one field of view, so it appears that "half" the Cygnus Loop is available at one time. With the filter and the monk's hood, I can just detect IC 1340 in between the two bright sections, but it is difficult and never held steady.

17.5 in. f/4.5 S=7 T=8 Excellent site

100×—With a 20 mm Erfle eyepiece and a UHC filter, the Veil is amazing. Only about one quarter of either loop can fit into the field of view and the scope must be scanned to see all that is available. NGC 6992 has loops and swirls of nebulosity that give a three-dimensional effect. There are other pieces to the Veil Nebula, most of them between the two main sections, much of what can be photographed in an 8 in. Schmidt Camera can be viewed by a persistent observer. This is the object on which the UHC filter does its best work.

36 in. f/5 S=7 T=9 Excellent site

165×+2 in. UHC—WOW!!! I am giving up trying to come up with superlatives about this object in the 36 in. It looks like twisted candy from a three-dimensional taffy pull machine. Some portions are mottled and others are smooth sections with stars involved of a wide variety of magnitudes. There are several places where you can see small pieces of nebula detached from the main body of the Veil. Either of the two main sections has so much detail to see; it takes about eight fields of view to see it all. Truly a unique and magnificent object. My finest view of any object at any time in any telescope.

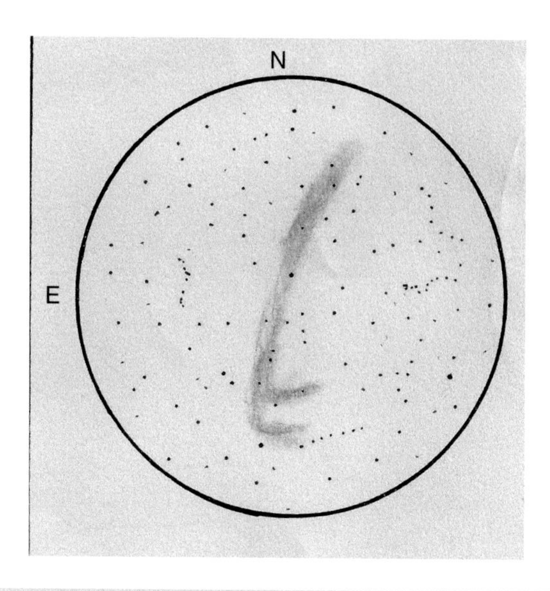

Fig. 11.25 6″ f/6; NGC 6992; FOV 1/2°; MAG 25× with UHC filter

Chapter 12

What Can Be Seen Within Planetary Nebulae?

A planetary nebula is formed at the end of the life of a star which has a mass about equal to that of our Sun. Any much more massive star cannot pass through this relatively gentle stage; it explodes in a shattering super-nova. A star much less massive than the Sun simply fades out slowly, like an ember from a campfire. As you look out into the Milky Way, over half of the stars you see will pass through this planetary nebula phase as their final curtain call.

Two centuries ago, a variety of small, greenish, disk-shaped nebulae were discovered by William Herschel; these objects reminded him of the size, color and shape of the planet Uranus. William Herschel remembered Uranus well, because his rise to prominence started with his discovery of that planet. So, he called this class of objects planetary nebulae. Actually, they have nothing in common with the planets, other than their appearance.

Many years of detailed study by astronomers have provided us with a fairly complete picture of how planetaries are formed and some answers about the myriad of shapes they present. Part of the fascination is that in about 5 billion years our lovely yellow Sun will also form one of these nebulae.

Right now and for many centuries to come the Sun is getting its energy by fusing hydrogen into helium. One day the hydrogen will start to be used up and helium will be fed into the fusion furnace at the center of the Sun. The helium will go through the fusion process to form heavier atoms: carbon, nitrogen, oxygen and many other elements. This process forms a dense core with layers of these elements around it, all surrounded by an outer atmosphere of gas. This is the "red giant" phase of a star's life cycle.

The hot core begins to use the energy of its radiation to push away the gas. This effect is called radiation pressure and it forms a large shell of material surrounding the

S.R. Coe, *Deep Sky Observing*, The Patrick Moore Practical Astronomy Series,
DOI 10.1007/978-3-319-22530-2_12

core of the red giant star. This stellar wind is removing the red giant's atmosphere at a fairly slow pace, about 10 km/s. This is called the "slow wind" phase of the formation of a planetary nebula. Because the outward force is greatest at the equator of the star, most of the material is ejected from there. However, as time goes by, more and more cool gas is flung into interstellar space and the gas cloud becomes somewhat barrel-shaped. Do not picture this as a smooth process; pulsations in the star's outer layer form thin and thick areas of gas and these will eventually be translated into bright and dark regions within the nebula.

When the mass of the remaining star is well below one solar mass, the slow wind stops. What is left is a very hot stellar core at the center of a huge envelope of relatively dense, cool gases that are moving slowly. Now the core starts to emit a fast-moving wind with speeds of the order of 1000–4000 km/s. Very quickly the fast-wind particles catch up with the slow-wind material and slam into them with tremendous energy. This is the "fast wind" phase.

The atoms accelerated by the fast wind collide with the atoms from the slow wind and electrons in these atoms are excited into higher orbital energy levels than before the collision. After a short period of time those atoms give up that energy as electromagnetic radiation. The wavelength of that radiation is specific to the atom that emits it. This process is called "collisional excitation" and it is one way in which a planetary nebula is caused to shine.

The other method that lights up all that gas is ionization. The characteristic blue–green color of many planetary nebulae is from the prominent radiation of doubly ionized oxygen (O III, "Oh three"). Two electrons are knocked free from an oxygen atom by the high-energy ultraviolet radiation emitted by the white dwarf star at the core. When two electrons return to the oxygen ion, their recombination energy is released in the visible spectrum, fortunately for deep-sky observers.

The two processes, collisional excitation and ionization, are both going on at the same time. Which is the dominant radiation producer is determined by the distance to the core star, the density and composition of the gases and how much dust is mixed in with this interaction.

There are many factors at play while forming the shape and size of a planetary nebula. The direction from the planetary to the observer is foremost. That barrel-shaped nebulosity can be edge-on to form a ring, tilted somewhat to form two rings or include bipolar jets exiting the poles of the star and shooting off into space. Many computer simulations have been done and they can be made to simulate many of the shapes actually seen in planetaries. Studies are pointing to planetaries having double stars or huge planets surrounding the central star; in some nebulae this might be a factor in directing the stellar winds and creating a variety of bizarre forms.

So, what can you see at the eyepiece? The short answer is: most anything! Obviously, after the previous discussion, you should look closely at the shape of the nebula. Try high power if the seeing will allow and look for light and dark areas within the nebulosity; also, try averted vision looking for a faint outer cloud of nebulosity. Much is written about seeing central stars in planetary nebulae and that is a fun game to try.

Because nebular filters were invented with these objects in mind, either put one into the eyepiece or blink with it, back and forth in front of the eyepiece. The blinking technique allows you to look for small nebulae in a rich star field. The nebula does not dim with the filter in place, but the stars do get fainter. Also, I use the blinking method with large, bright nebulae looking for changes in the nebulosity as the filter is added and then taken away.

Planetary nebulae are some of my favorite objects in the sky. They present a wide variety of sizes and shapes. Many of the most famous have a high surface brightness and therefore are easy to find and show off to a novice. Spend some time drawing them and you can compare the smorgasbord of nebulae all rendered by your own hand.

Vorontsov–Velyaminov types for planetary nebulae	
1 Stellar	
2 Smooth disk	a: brighter center
	b: uniform brightness
	c: traces of ring structure
3 Irregular disk	a: very irregular brightness distribution
	b: traces of ring structure
4 Ring structure	
5 Irregular form similar to diffuse nebula	
6 Anomalous form, no regular structure	

Note: Some very complex forms may combine two types

Object	**NGC 1535**
Other names	**H IV 26**
Type	**PLNNB**
Mag	**10.4**
Size	**20″ × 17″**
Class	**4(2c)**
Central star	**12.1**
Surface brightness	**7.2**
Constellation	**Eri**
RA	**04 14.2**
Dec	**−12 44**
Tirion	**11**
U2000	**268**
Description	**vB, S, R, pS, vsbM, r**

6 in. f/6 S=6 T=8 Excellent site

40× Pretty faint, small, bright middle, just seen as a round nonstellar disk at low power.

135×—Now this planetary is pretty faint, pretty small, little elongated and shows a stellar nucleus. Averted vision doubles its size. There is no color seen, it is a grey disk with white stellar core.

13 in. f/5.6 S=8 T=8 Excellent site

100×—Bright, small, round, seen as a disk easily, needs more power.

330×—Now it is pretty large, round, and somewhat brighter in the middle, the central star easy. There are two concentric rings, one bright, elongated and near the center, the other ring dimmer and larger. The fainter large ring is what gives this nebula its round appearance at lower powers. This nebula is light blue at all magnifications.

36 in. f/5 S=7 T=9 Excellent site

390×—Shows a bulls-eye of nebulosity, with the inner ring a light-green oval and the faint outer fringe a light pink. The central star is easy and has a very thin dark region surrounding it. There is a very faint star involved at the edge of the nebula.

A nice planetary to break up all those faint galaxies in Eridanus (See Fig. 12.1).

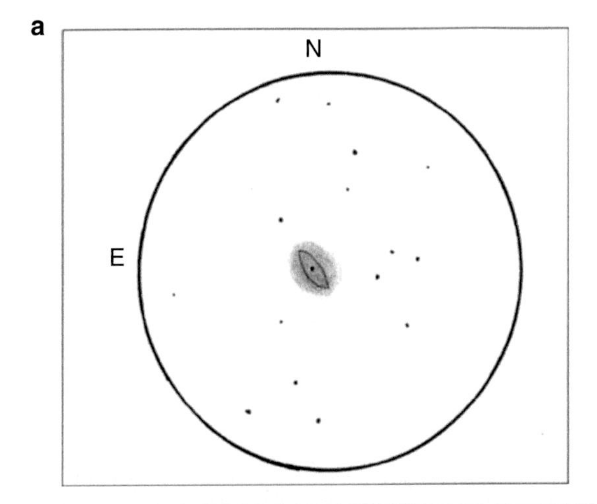

Fig. 12.1 (**a**) 13″ f/5/6′ NGC 1535; FOV 10′; MAG 330×. (**b**) 12″ Newtonian; *Photo: Chris Schur*

Object	NGC 2371
Other names	H II 316
Type	PLNNB
Mag	13.0
Size	74″ × 54″
Class	3a + 6
Central star	12.5
Surface brightness	10.5
Constellation	Gem
RA	07 25.6
Dec	+29 29
Tirion	5
U2000	100
Description	B, S, R, bMN, p of Dneb

This object got assigned two NGC numbers, NGC 2371 and 2372.

4 in. f/8 S=6 T=7 Very good site

90×—Faint, small, elongated 1.5 × 1 in PA 45°. Averted vision shows some better contrast but this nebula is not much with the TV 102 telescope.

13 in. f/5.6 S=8 T=8 Excellent site

100×—Bright, pretty small, elongated 1.5 × 1 in PA 135, double nature seen 10 % of the time with averted vision.

330×—Great view. The double object is obvious, the two nebulae split with a thin dark lane between them. The west lobe is brighter with a bright, non-stellar area within it. There is a faint 13th mag central star just between the two lobes. There is a thin dark lane between the two sections. This nebula is gray in color at all powers. The UHC filter shows no new detail.

36 in. f/5 S=7 T=9 McDonald Obs. Excellent site

390×—NGC 2371 and 2372 are gray in color. There are two sections and dim nebulosity between the lobes, with a central star between the two parts. The western portion is brighter and somewhat larger.

Averted vision shows a very faint outer loop of nebulosity about twice the size of the bright nebulae and most prominent on the north side.

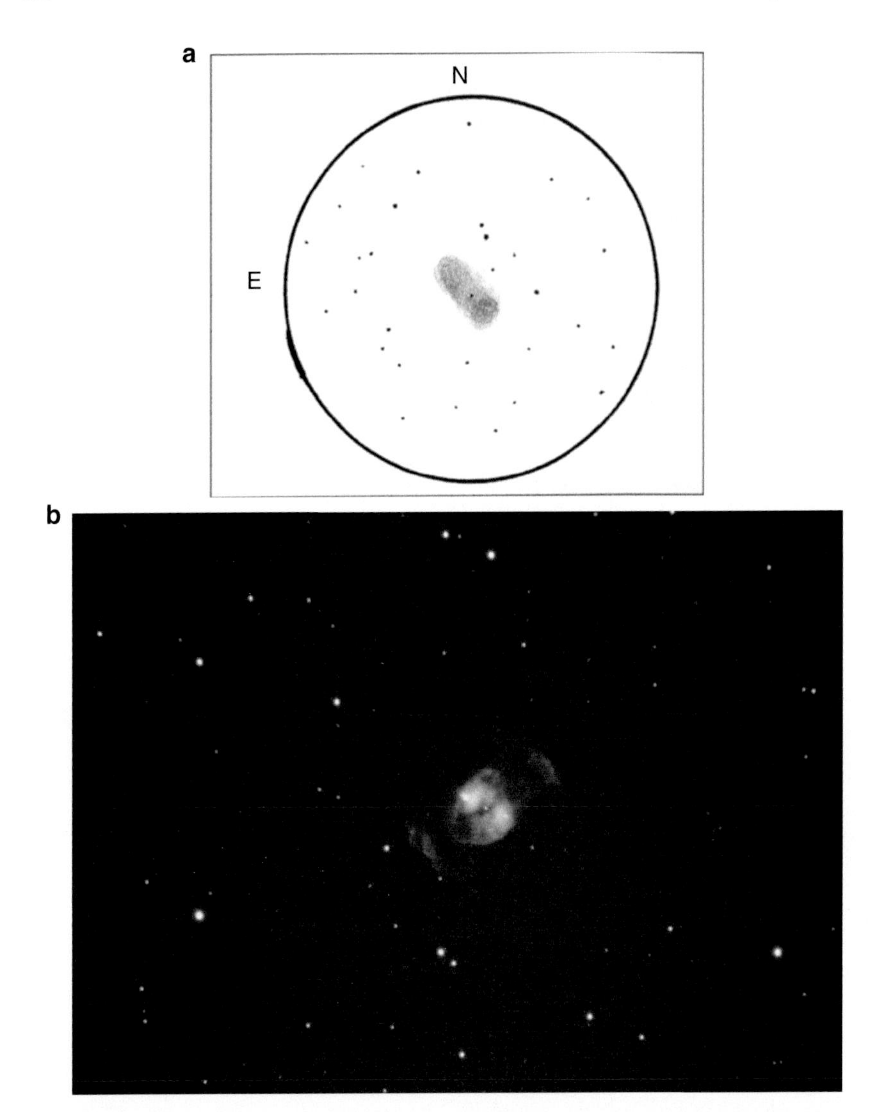

Fig. 12.2 (**a**) 13″ f/5.6′ NGC 2371; FOV 20′; MAG 220×. (**b**) C14 at f7,6 Atik 314L+; *Photo: Parijat Singh*

Object	**NGC 2392**
Other names	**H IV 45**
Type	**PLNNB**
Mag	**8.6**
Size	**47″ × 43″**
Class	**3b(3b)**
Central star	**10.6**
Surface brightness	**6.8**

Constellation	**Gem**
RA	**07 29.2**
Dec	**+20 55**
Tirion	**5**
U2000	**139**
Description	**B, S, R, *9M, *8 nf 100″**

The Eskimo Nebula got its name from the fact that dark markings within the nebulosity look like a human face with a parka around it, formed by a faint outer gas cloud. Burnham says it reminds him of W.C. Fields, central star as a big nose and droopy eyebrows. I can see that.

6 in. f/8 S=7 T=7 Good site

135×—Bright, pretty small, round, the central star is easy; there are two levels of brightness within the nebula with the area nearest the central star the brightest. Averted vision makes it larger in size, but makes it a pretty smooth disk. The disk is very light green.

180×—the two levels of brightness are easier, there is a hint of the dark markings around the central star. 255×—the dark markings that create the "Eskimo" face are much more prominent at this power.

17.5 in. f/4.5 S=5 T=6 Mediocre site

100×—Bright, large, round, central star easy, annular disk fills in with averted vision.

250×—Central star still held steady, now two rings of bright nebulosity around star. Light green in color.

13 in. f/5.6 S=8 T=8 Excellent site

100×—Bright, pretty large, round, bright nucleus, even at low power a nice aqua disk with obvious central star. Averted vision is almost startling—direct vision is star only, averted vision is big, bright disk. There is a bright star, about 8th mag to the north and an 11th mag star just to the northwest of the nebula.

440×—A great view. The central star is still obvious and stands out from the nebula nicely. The disk is BIG and there are several dark markings. Two are curved around the central star from the "face" of the Eskimo.

There is a fainter outer rim of nebulosity that is only seen with averted vision. A terrific view of a terrific object, lots of fine detail.

25 in. f/5 S=7 T=9 Excellent site

250×—Beautiful, dark-green color, dark lane around the central star, several small dark markings within nebula. Averted vision shows a fuzzy annulus surrounding the entire nebula. The "face" of the Eskimo is obvious once you orient your eye correctly and see the markings as a human face with a furry hood.

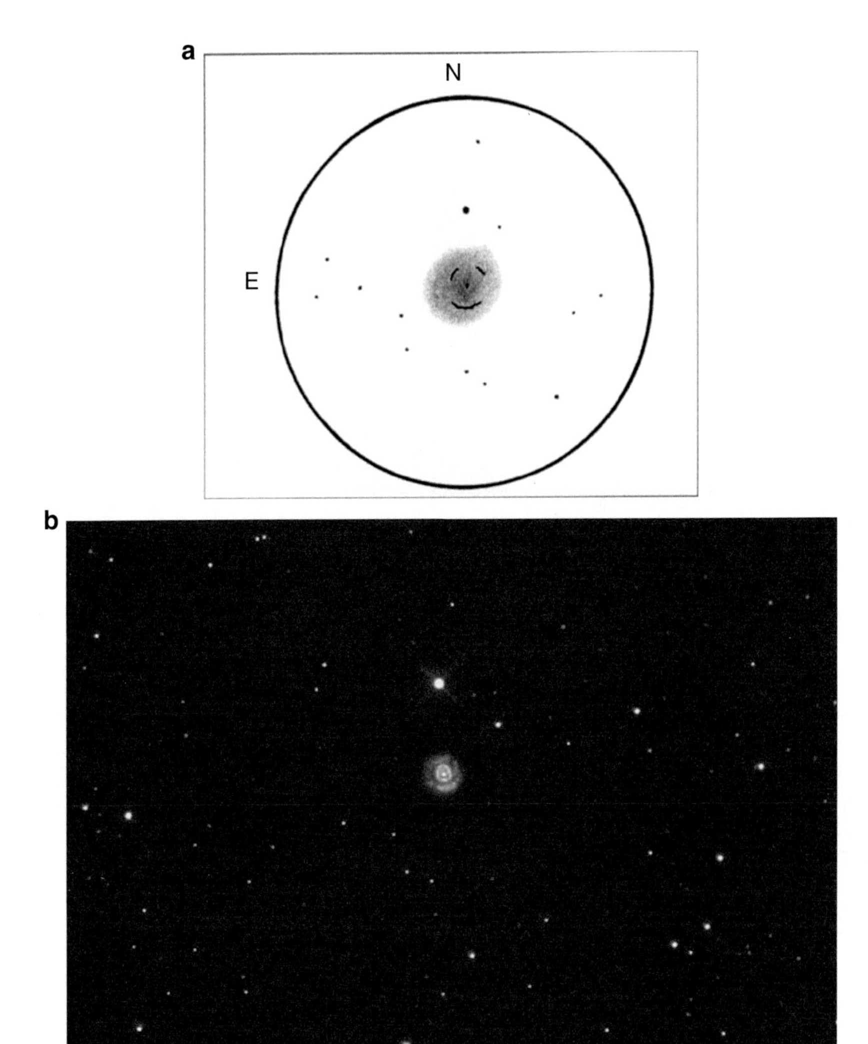

Fig. 12.3 (**a**) 13″ f/5.6; NGC 2392; FOV 7′; MAG 440×. (**b**) 12″ Newtonian; *Photo: Chris Schur*

Object	**NGC 2438**
Other names	**H IV 39**
Type	**PLNNB**
Mag	**11.0**
Size	**65″**
Class	**4(2)**
Central star	**17.5**
Surface brightness	**11.7**
Constellation	**PUP**

RA	**07 41.8**
Dec	**−14 44**
Tirion	**12**
U2000	**274**
Description	**pB, pS, vlE, r**

A unique object, a very nice planetary nebula at the edge of an open star cluster (M46).

6 in.	f/8	S=6 T=7	Good site

90×—Easy to see as non-stellar this planetary is pretty bright, pretty small and somewhat elongated.

180×—the nebula is elongated 1.8×1 in a wedge shape. The higher power shows two stars involved, one is at the tip of the wedge or horseshoe shape.

13 in.	f/5.6	S=7 T=8	Excellent site

100×—Bright, large, slightly elongated 1.2×1 in PA 135, the central star is easy, even at this low power.

440×—Brings out two dimmer stars involved in the nebula. Also at the higher power the shape of the planetary is seen to be an incomplete ring, somewhat like a horseshoe. This bright rim is about 270° around and is brightest on the north side. I have always seen this planetary nebula as light green in color.

Notice that the data from the SAC database says that the central star in NGC 2438 is 17.5 magnitude. Therefore the star I had thought of as the "central" star is not the hot white dwarf that lights up the nebula. I assume that the stars in the planetary are either just line-of-sight stars in front of the nebula or are stars that have been enveloped by the expanding gas cloud.

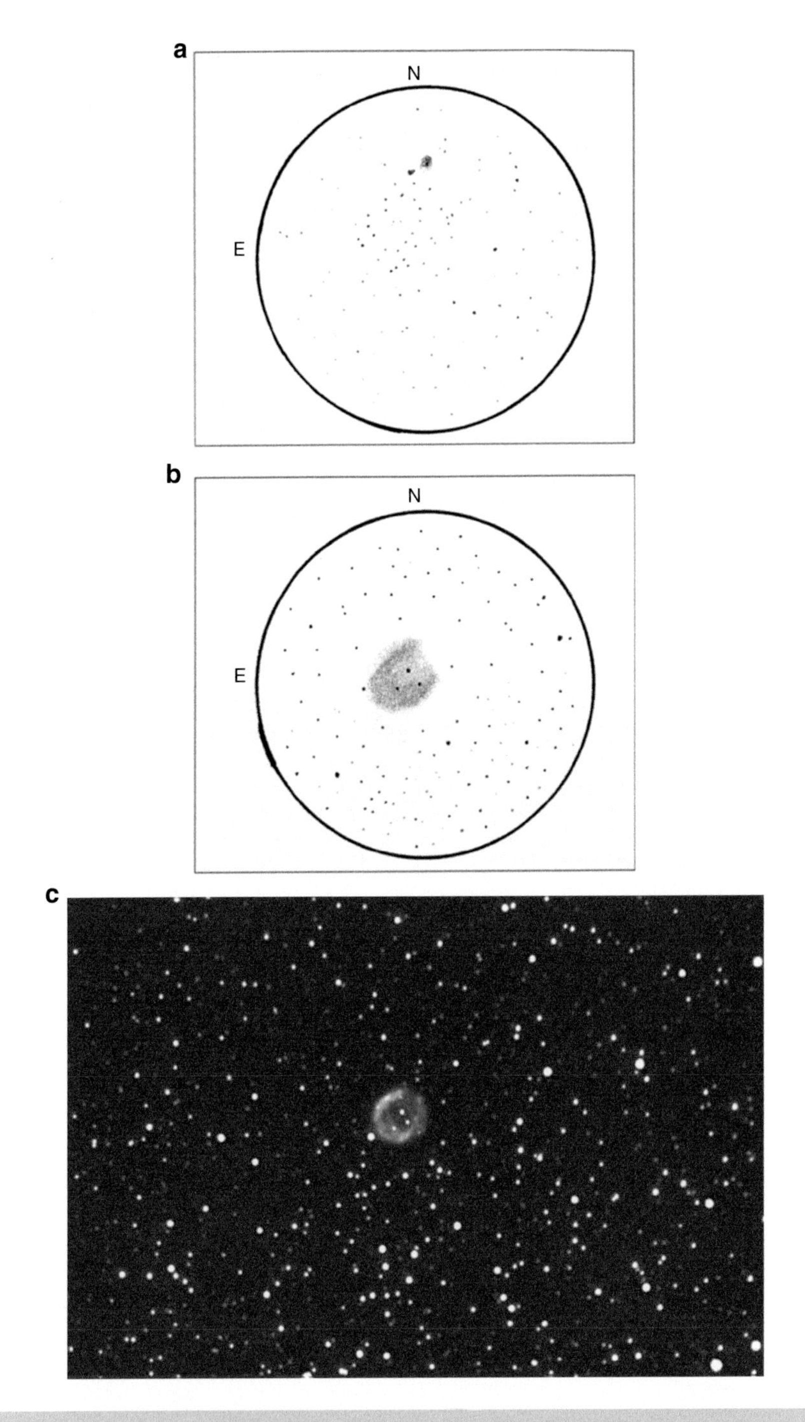

Fig. 12.4 (**a**) 13″ f/5.6; NGC 2438 and M46; FOV 30′; MAG 100× with UHC filter. (**b**) 25″ f/5; NGC 2438; FOV 5′; MAG 400×. (**c**) 12″ Newtonian; *Photo: Chris Schur*

Object	NGC 2440
Other names	**H IV 64**
Type	**PLNNB**
Mag	**11.5**
Size	**54″ × 20″**
Class	**5(3)**
Central star	**17.5**
Surface brightness	**6.4**
Constellation	**PUP**
RA	**07 41.9**
Dec	**−18 13**
Tirion	**12**
U2000	**319**
Description	**cB, not v well def**

13 in. f/5.6 S = 7 T = 8 Excellent site

100×—Pretty bright, pretty large, elongated 1.5 × 1 in PA 45.

220×—Pretty suddenly much brighter middle, edges fuzzy, not well defined central bright spot is never stellar.

330×—Amazing detail in bright central section, looks turbulent, several bright areas interconnected; averted vision will show an outer band of nebulosity that doubles the size of the entire nebula. This nebula is lime-green at all powers.

I have not had a chance to point a large scope at this object, but I am certain that on a clear night it would show some fascinating detail at high power. This is on my observing list (See Fig. 12.5).

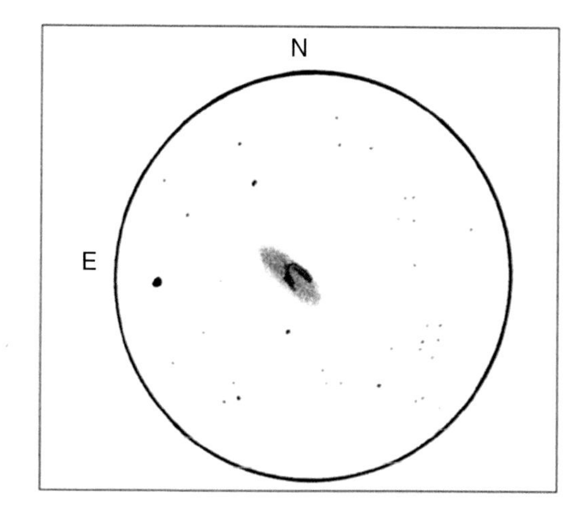

Fig. 12.5 13″ f/5.6; NGC 2440; FOV 10′; MAG 330×

Object	**NGC 3242**
Other names	**H IV 27**
Type	**PLNNB**
Mag	**8.6**
Size	**40″ × 35″**
Class	**4(3b)**
Central star	**12.3**
Surface brightness	**5.4**
Constellation	**Hya**
RA	**10 24.8**
Dec	**–18 38**
Tirion	**13**
U2000	**325**
Description	**! vB, lE 147,45″ d, blue**

I know that this planetary was christened "Ghost of Jupiter" by Admiral Smyth, but it certainly has features that look more like an eye looking back at you. In the US the CBS network has used an eye as their logo for decades, so we call this the "CBS Eye Nebula".

6 in.	f/8	S=6 T=7	Good site

90×—Bright, pretty large and little elongated 1.2×1. This planetary shows an electric aqua color even in only 6 in. of aperture.
190×—displays the "CBS eye" feature in low contrast but it is seen with direct vision.
255×—I consider 255× the highest useful power with the 6″ f/8 refractor. The interior detail is more easily seen at this magnification. It does indeed look like a cosmic eye looking back at me. The central star winked at me twice, but was never held steady.

13 in.	f/5.6	S=7 T=8	Excellent site

100×—Bright, pretty large, little elongated 1.2×1 PA 135; the central star comes and goes with the seeing, the nebula is aqua.
220×—"CBS eye" detail is obvious, a dark background and bright oval that encompassed the central star. Very light green. Adding the UHC filter makes the central star disappear.
440×—Best view, several bright knots to southeast of central star, bright spots within oval that surrounds the central star. High power makes the color gray, not green, but high power brings out most detail, including central star and bright knots in disk.

36 in.	f/5	S=7 T=8 Texas Star Party	Excellent site

300×—Bright, large, lots of detail, three levels of brightness, central star held constant; there is a dark ring around central star. Beyond the bright nebula is a very dim outer section, only seen with averted vision. Great view, medium-green color at all times. A few bright knots in the brighter nebulosity that forms the CBS eye.

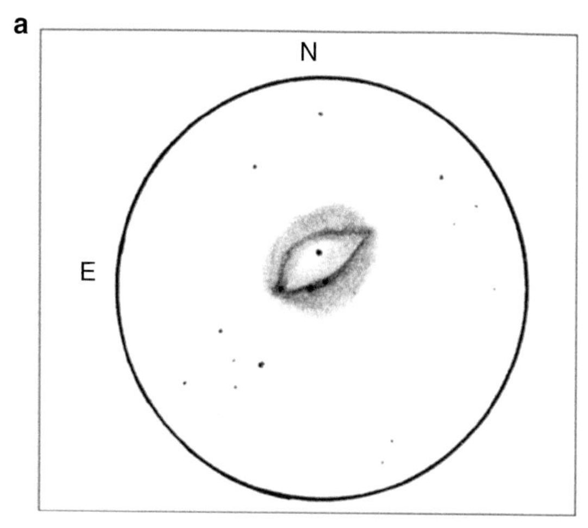

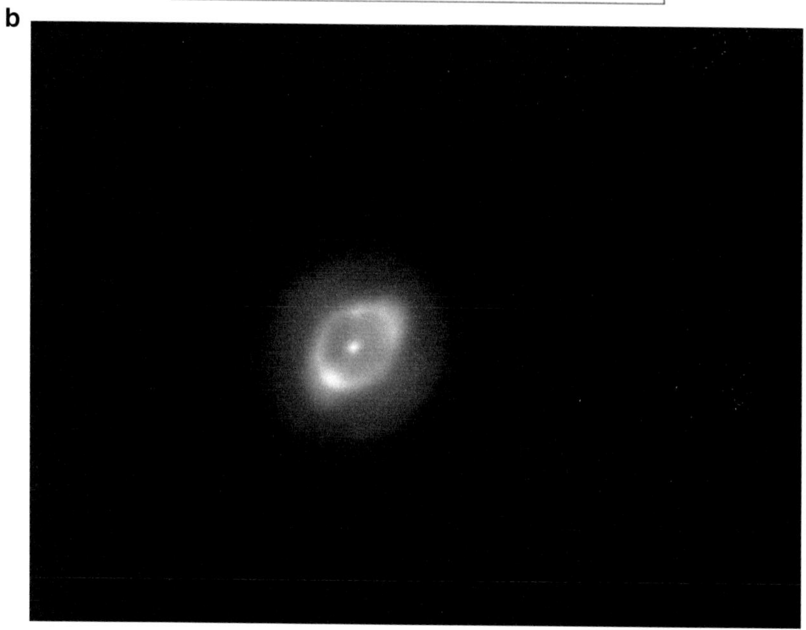

Fig. 12.6 (**a**) 36″ f/5; NGC 3242; FOV 7′; MAG 350×. (**b**) AT10RC QSI 640wsg; *Photo: John Dwyer*

Object	NGC 6210
Other names	**PK43+37.1**
Type	**PLNNB**
Mag	**9.7**
Size	**20″×13″**
Class	**2(3b)**
Central star	**12.5**
Surface brightness	**5.9**
Constellation	**Her**
RA	**16 44.5**
Dec	**+23 49**
Tirion	**8**
U2000	**156**
Description	**vB, vS, R, disc**

13 in. f/5.6 S=6 T=7 Good site

100×—Bright, small, a little brighter in the middle, nebular disk easily seen as non-stellar even at low power, about three times the size of the seeing disk. A lovely blue–green color.

330×—Pretty bright, still pretty small, bright middle with a stellar nucleus, little elongated 1.2×1 in PA 90. The disk grows with averted vision. The color was better at 100×; it is a washed-out gray at high power.

13 in. f/5.6 S=8 T=10 Superior site

100×—Disk is approximately five times the size of the seeing disk of a star and GREEN, unmistakably Kelly Green, pretty bright, small; the central star is seen 20 % of the time.

220×—Disk is little elongated 1.5×1 in PA 135; the central star is held steady with direct vision.

330×—Averted vision somewhat dilutes the color to gray–green, but it shows a faint outer shell that extends from the PN in all directions. I observed a nearby star of equal magnitude and it did not show this faint shell, so it is a feature of the planetary. Wait for those rare nights when the sky is clear and steady, then you can really use high magnification to see new features in objects you might have observed often on good but not spectacular evenings.

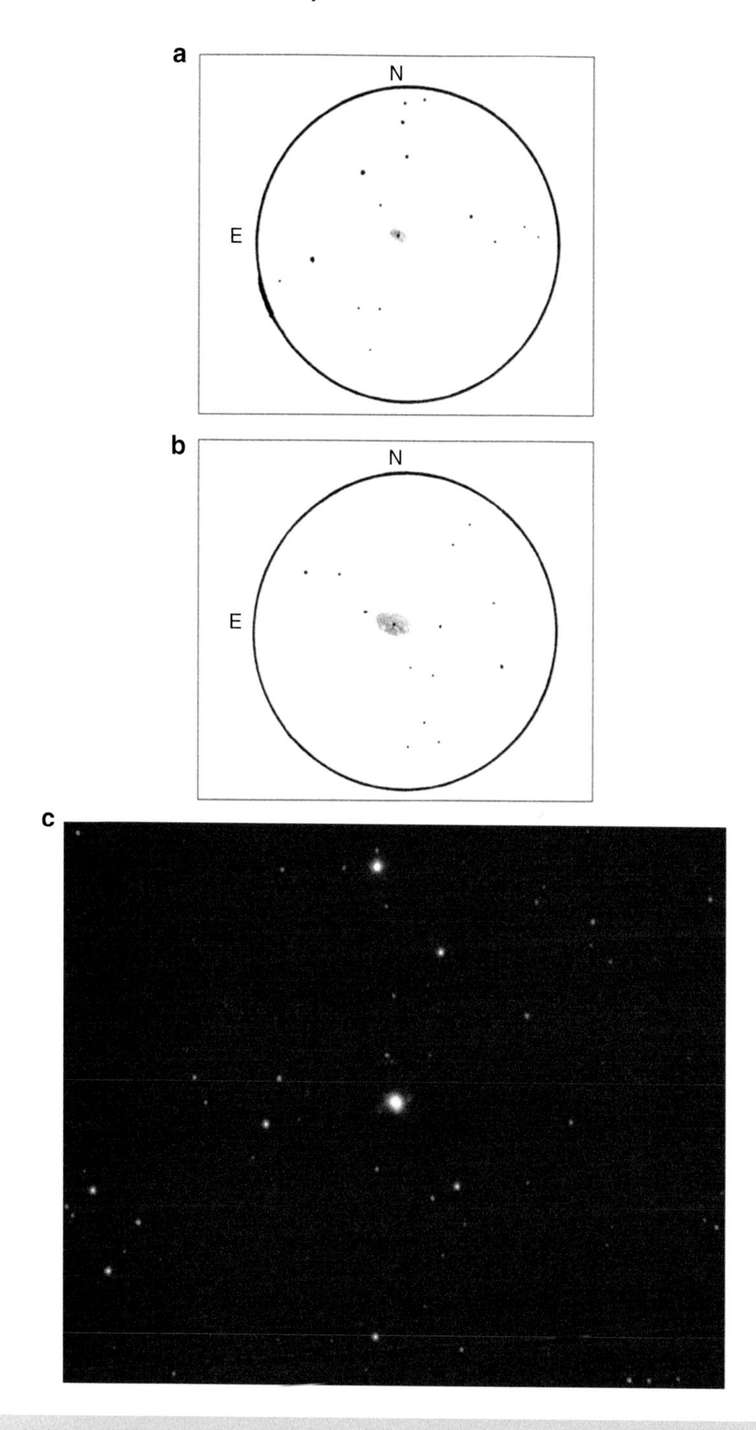

Fig. 12.7 (**a**) 17.5″ f/4.5; NGC 6210; FOV 20′; MAG 130×. (**b**) 13″ f/5.6; NGC 6210; FOV 10′; MAG 330×. (**c**) 12 in. LX 200 Canon T3i; *Photo: Terry Riopka*

Object	**NGC 6302**
Other names	**PK349 + 1.1**
Type	**PLNNB**
Mag	**12.8**
Size	**2′ × 1′**
Class	**6**
Central star	**16.6**
Surface brightness	**9.0**
Constellation	**Sco**
RA	**17 13.7**
Dec	**−37 06**
Tirion	**22**
U2000	**376**
Description	**pB, E pf**

A bizarre shape gives this object the nickname the Bug Nebula, as in a bug that has been squashed under a swatter. Notice that the Vorontsov–Velyaminov type for this object is "6"; that always denotes a strange shape for a planetary and this jumbled, non-symmetrical glow is certainly strange.

6 in. f/6 S = 7 T = 7 Excellent site

65×—Pretty faint, small, elongated 2 × 1, very much brighter in the middle. With the 6 in. it only shows a star with a little fuzz around it. Moderately difficult to find in a rich field of view.

13 in. f/5.6 S = 7 T = 8 Excellent site

150×—Easy to find at this power, pretty bright, pretty large, elongated 2 × 1 in PA 75. Central "star" is not stellar, about three times the size of the Airy disk of nearby stars. There is a dark lane to the west of center.

Ken Reeve's 20 in. f/5 same night

285×—There are two dark lanes on the west side and one to the east of the central star. These dark markings chop the nebulosity into unequal parts. There is an obvious bright spot within the western section, it is also much larger than the Airy disk. Averted vision shows some outer, faint nebulosity to the north and south of the main bright nebula. Ken's O III filter brings out some of the faint outer nebulosity but does not add any new detail to the central section.

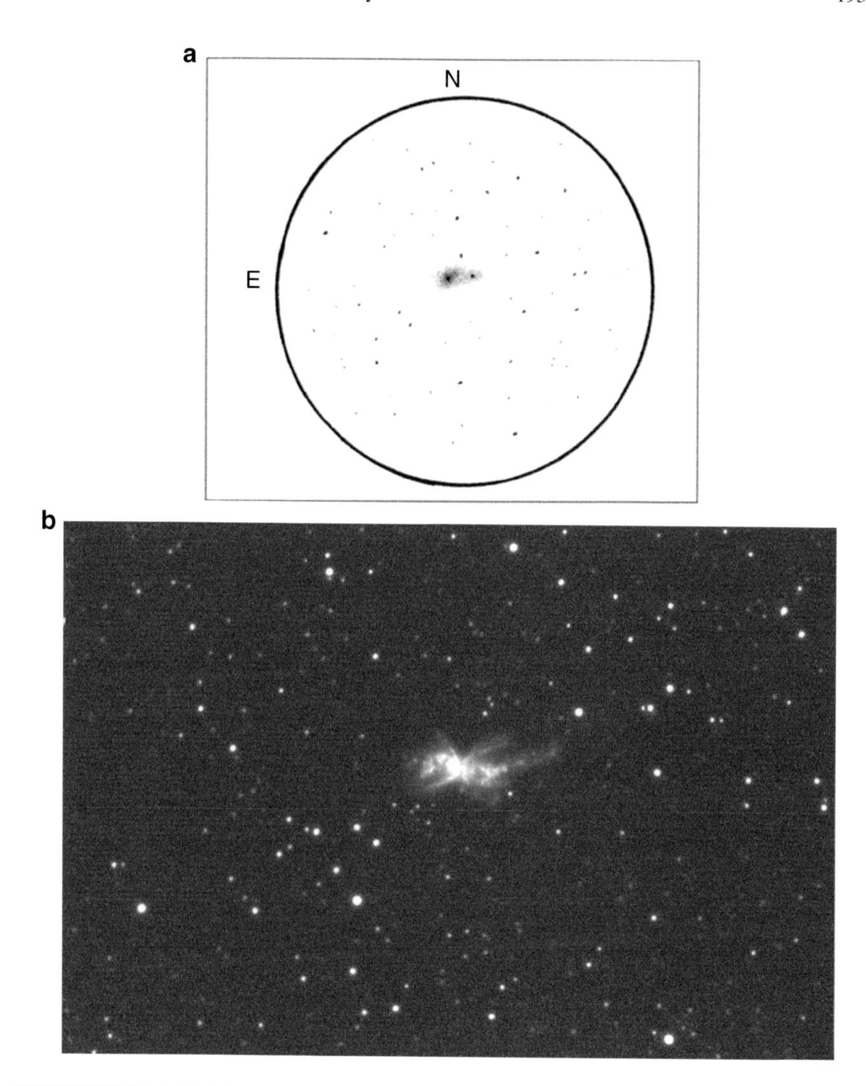

Fig. 12.8 (**a**) 17.5″ f/4.5; NGC 6302; FOV 10′; MAG 320×. (**b**) 12″ Newtonian; *Photo: Chris Schur*

Object	**NGC 6369**
Other names	**PK2+5.1, H IV 11**
Type	**PLNNB**
Mag	**11.0**
Size	**30″×29″**
Class	**4(2)**
Central star	**15.1**
Surface brightness	**10.5**
Constellation	**Oph**
RA	**17 29.3**
Dec	**−23 46**
Tirion	**22**
U2000	**338**
Description	**Annular, pB, S, R**

This object is located within the "bowl" of the dark Pipe Nebula, so there are few field stars.

6 in.	f/6	S=6 T=6	Good site

40×—Not seen as anything other than a star.
100×—Extremely small, faint, round. Averted vision helps to determine that this is an object larger than the seeing disk of the nearby stars.

13 in.	f/5.6	S=6 T=7	Good site

100×—Pretty bright, small, round, much brighter in the middle, easy to see as a non-stellar disk.
220×—Pretty bright, pretty large, little elongated 1.2×1 PA 90. This planetary is much brighter on the north side and is annular with averted vision. It is light green at all powers. No star was seen within the annulus.
There is a bright outer lip that forms a horseshoe shape at the limb of the nebulosity.

36 in.	f/5	S=7 T=7 Texas star party	Excellent site

300×—Bright, pretty large at this high power, somewhat elongated.
The nebula is a light green in color and the bright outer annulus is an incomplete horseshoe. The central section of the annulus has a very faint nebulosity with in it that fills in the annulus; it can just be seen with averted vision. The central star is seen 20 % of the time. Adding a UHC filter blanks out the star but the central nebulosity is now much easier to see without averted vision.

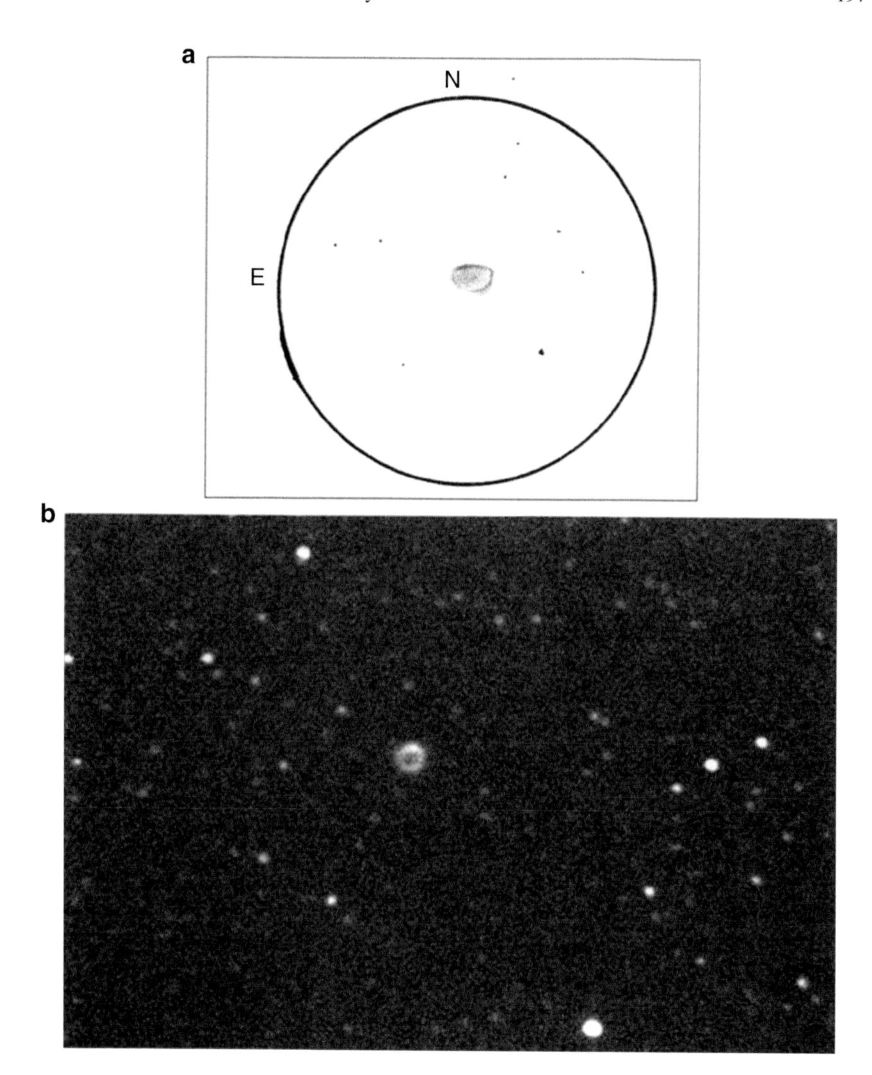

Fig. 12.9 (**a**) 13″ f/5.6; NGC 6369; FOV 20′; MAG 220×. (**b**) 12″ LX 200 at f/7 ST7E CCD; 6 min exposure; *Photo: Larry E. Robinson*

Object	**NGC 6445**
Other names	**PK8+3.1, H II 586**
Type	**PLNNB**
Mag	**13.0**
Size	**35″×30″**
Class	**3b(3)**
Central star	**19.0**
Surface brightness	**9.6**

(continued

(continued

Constellation	**Sgr**
RA	**17 49.2**
Dec	**–20 01**
Tirion	**22**
U2000	**338**
Description	**pB, pS, R, gbM, r, *15 np**

6 in.	f/6	S=6 T=6	Good site

40×—Pretty faint, pretty small, round, not brighter in the middle.
It is "above" a globular of about the same brightness (NGC 6440), but the globular is about twice the size of this planetary.
100×—Still pretty faint, but now has some size. It is elongated 2×1 in PA 135. Tough to determine PA with a Dobsonian scope.

13 in.	f/5.6	S=7 T=7	Excellent site

150×—Pretty bright, pretty large, not brighter in the middle, elongated 1.8×1 in PA 165. It appears as an elongated box shape with a bright outer rim. With averted vision there is a thin dark lane down the middle. This nebula has a high surface brightness and is gray in color. A nice, easily split, white-and-blue double star is to the east of this planetary nebula.
Ken Reeve's 20 in. f/5 same night
285×—Light green in color, a bizarre shape, two rounded squares of nebulosity with a dark lane in the middle. The outer rim of the nebula is brighter than anything inside it. With averted vision a very faint loop of nebulosity extends away from the bright section toward the south. A unique object.

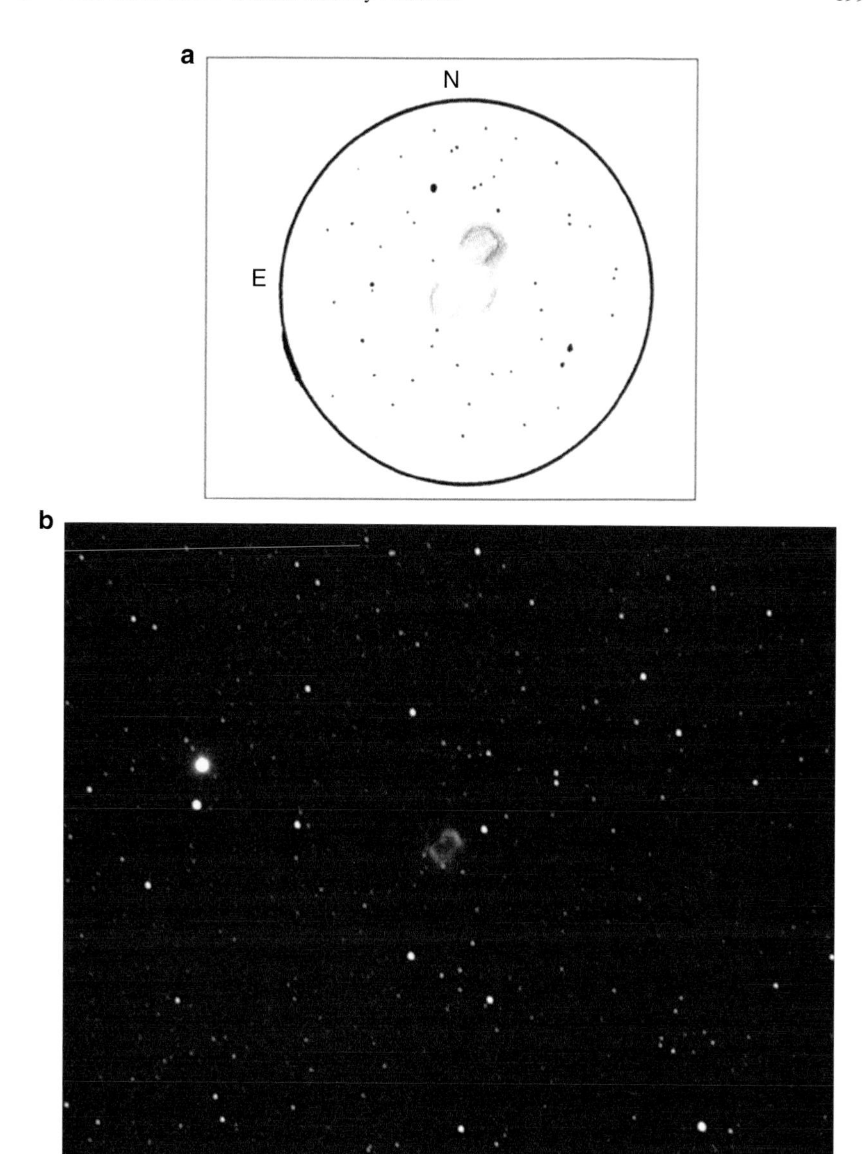

Fig. 12.10 (**a**) 20″ f/5; NGC 6445; FOV 12′; MAG 300×. (**b**) 12 in. LX 200 Canon T3i; *Photo: Terry Riopka*

Object	**NGC 6572**
Other names	**PK34+11.1**
Type	**PLNNB**
Mag	**8.0**
Size	**15″×12″**
Class	**2a**
Central star	**12.0**
Surface brightness	**4.3**
Constellation	**Oph**
RA	**18 12.1**
Dec	**+06 51**
Tirion	**15**
U2000	**204**
Description	**vB, vS, R, I haxy**

Don't let the small size of this luminescent gem stop you from putting it on your observing list; it is worth it. In every scope I have ever owned, from an 8 in. to an 18 in. this is the greenest nebula I have ever seen! This little planetary is as green as an Irishman's coat on St Patrick's Day.

4 in.	f/8	S=8 T=8	Excellent site

120×—Pretty bright, very small, little elongated 1.2×1 in PA 90° (east–west). The central star winked at me about 5 % of the time.

13 in.	f/6	S=6 T=5/10	Mediocre site

100×—Bright, small, elongated 1.5×1 in PA 135, immediately obvious, a beautiful Easter egg afloat in the Milky Way.

220×—Averted vision shows a dim outer haze; there is a bright center that is about 4×5 arcsec in size, an outer hazy area twice that size. The central star glimpsed 20 % of the time.

330×—Outer haze is obvious. Central star is now seen about 50 % of the time and the star is a blue–white color. The nebula is still a beautiful blue–green color.

36 in.	f/5	S=7 T=9 McDonald Obs.	Excellent site

250×—A glorious green luminescent color; the elongated bright inner portion of the nebula contains an obvious central star and there is a dim outer section or envelope larger than the main bright body.

Averted vision will double the size of this nebula.

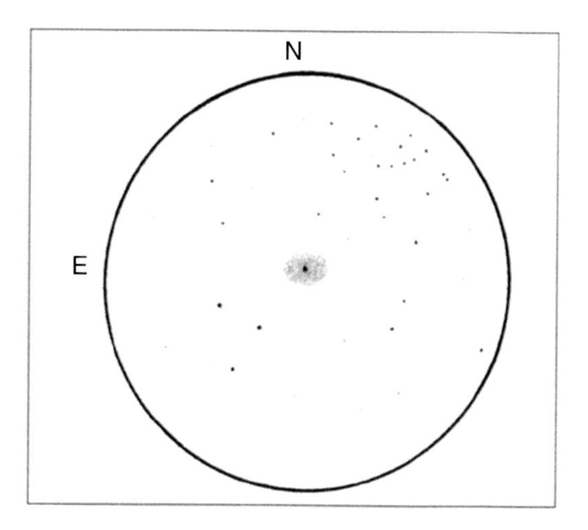

Fig. 12.11 13″ f/5.6; NGC 6572; FOV 20′; MAG 220×

Object	**M57**
Other names	**NGC 6720**
Type	**PLNNB**
Mag	**9.4**
Size	**86″×62″**
Class	**4(3)**
Central star	**15 var**
Surface brightness	**9.3**
Constellation	**Lyr**
RA	**18 53.6**
Dec	**+33 02**
Tirion	**8**
U2000	**117**
Description	**Ring neb, B, pL, cE**

Of all the objects I have ever seen I have probably observed the Ring Nebula with the widest variety of telescopes and magnifications. It is easy to find and has a high surface brightness, so that lots of people view and enjoy this famous planetary. The game of "Can you see the central star?" is played every clear summer night. John Herschel commented on the fact that the central section of the Ring looks filled in, he said, "like gauze across a hoop" (See Fig. 12.12).

6 in.	f/6	S=5 T=6	Good site

65× Pretty bright, small, a delicate and somewhat elongated donut afloat in the Milky Way. No stars are very near the Ring—except one lonely, 13th mag star less than 1 arcmin to the east. With a small aperture, the Ring resides in a dark area for about 10 arcmin all around.

13 in. f/5.6 S=5 T=6 Good site

11×80 finder—Just seen as 9th mag star.

60×—Annular glow even at low power—it is the Ring Nebula all the time!

150×—Great view, bright, large, elongated 1.5×1 in PA 75 a lovely greenish donut is a nice field of faint stars. The "gauze across a hoop" effect is obvious.

220×—Color is duller at higher power—a gray–green—but more detail is easier; the filled-in or "gauze" effect is seen always with direct vision, but is more evident with averted vision. Southeast and northwest quadrants are brighter than the rest of the Ring. The UHC filter enhances the contrast of the central fill in effect somewhat.

Color filters: light-green and light-blue filters don't seem to do much to the Ring itself, the stars nearby taking on the color of the filter. The light-orange filter almost makes the Ring disappear; it becomes just a faint glow.

13 in. f/5.6 S=8 T=10 Superior site

600×—At this high power the Ring is about half of the field of view. The color of the bright annulus is very diluted, a pale gray–green. The central star is seen about 5 % of the time, flickering on rarely. There is just a hint of some nebulosity outside the annulus, seen only with averted vision and the UHC filter in place. This could be one of those things I know is there, so I expect to see it. That outer nebulosity is at the limit of the 13 in. scope on a night with perfect transparency. However, I consistently see on extremely faint glow that extends beyond the Ring about 10 arcmin on all sides.

36 in. f/5 S=6 T=8 Excellent site
McDonald Obs.

300×—Medium-green, pretty faint star outside ring is easy, central star is held about 20 % of the time. The "gauze across a hoop" effect is obvious and held steady all the time. The slender fingers of nebulosity that point inward from the Ring toward the central star can be observed in moments of good seeing. Adding the O III filter shows some of the nebulosity beyond the Ring annulus. It is low in contrast, even with the filter, but I can hold it constantly with averted vision. A fabulous view of one of the most famous deep-sky goodies in the heavens, WOW!

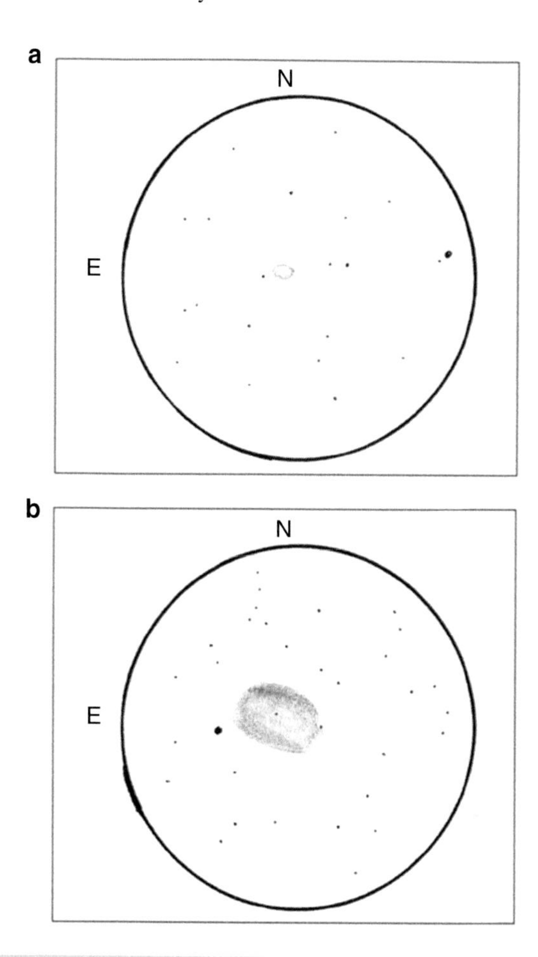

Fig. 12.12 (**a**) 6″ f/6; M 57; FOV 50′; MAG 65× (**b**) 13″ f/5.6′; M 57; FOV 10′; MAG 330×. (**c**) 36″ f/5; M 57; FOV 7′; MAG 600× (**d**) M 57; 12″ f/5. *Photo: Chris Schur*

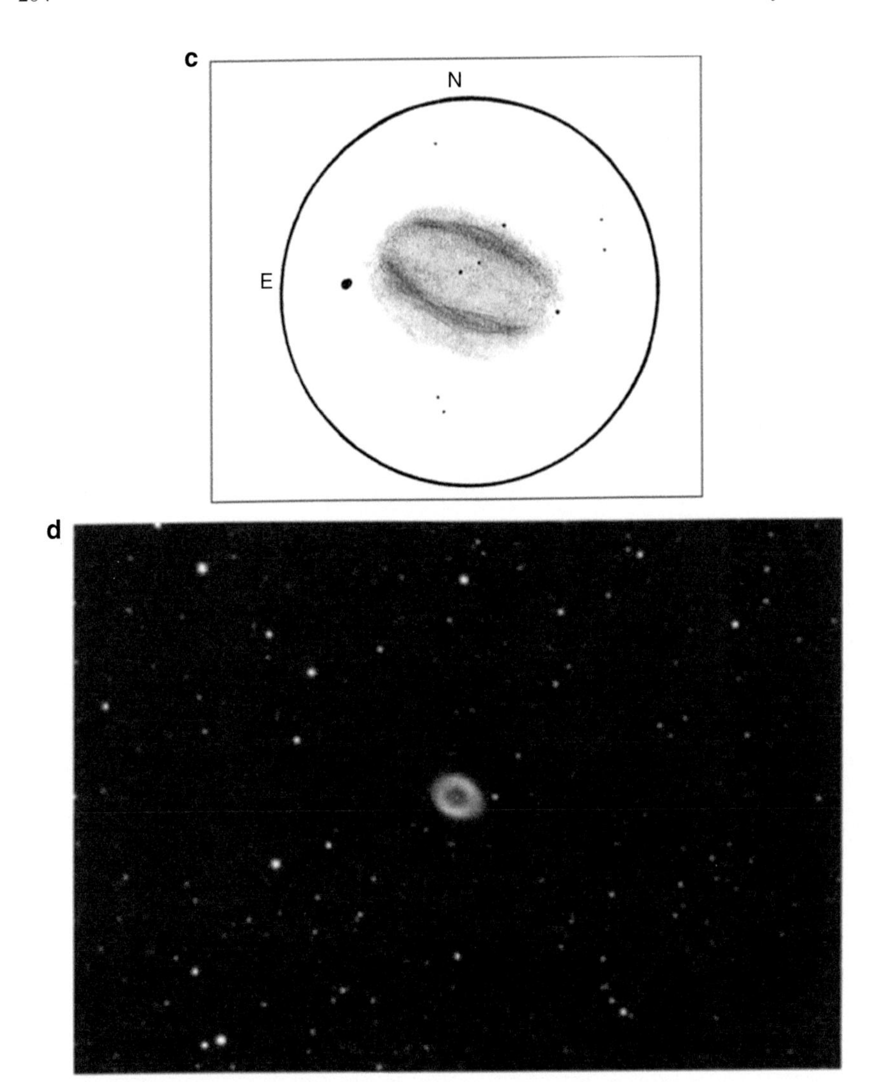

Fig. 12.12 (continued)

Object	NGC 6781
Other names	H III 743
Type	PLNNB
Mag	11.8
Size	111″ × 109″
Class	3b(3)
Central star	16.8
Surface brightness	12.8
Constellation	Aql
RA	19 18.5
Dec	+06 32
Tirion	16
U2000	206
Description	F, L, R, vsbM disc

6 in. f/6 S=7 T=7 Good site

40×—Easy to find, a pretty small disk.

100×—Faint, round, pretty large, not brighter in the middle, one star involved comes and goes with seeing, brighter on east side.

13 in. f/5.6 S=8 T=8 Excellent site

100×—Bright, large, somewhat elongated 1.5×1 PA 90. It is immediately obvious without the UHC filter. This planetary is shaped like the gibbous moon with the south side brighter in an arc.

There is one star involved that stands out very nicely. The UHC filter raises the contrast somewhat.

36 in. f/5 S=7 T=9 Excellent site
McDonald Obs.

300×—Bright, pretty large at high power, a gray disk. There is a curved bright region on the eastern side that makes the entire nebula appear like a crescent Moon filled in with Earthshine. There are two pretty faint stars, both right at the edge of the nebulosity, one on the north side and another on the west (See Fig. 12.13).

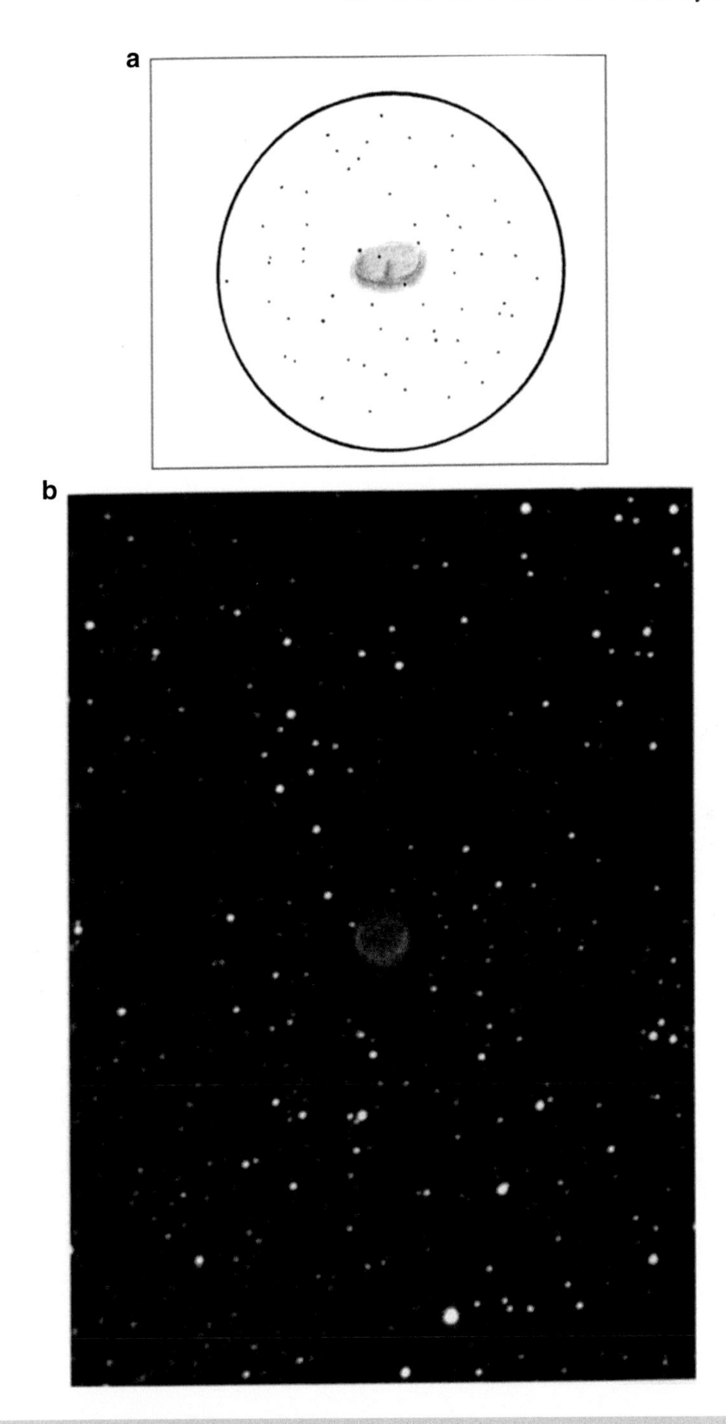

Fig. 12.13 (**a**) 13″ f/5.6; NGC 6781; FOV 10′; MAG 330×. (**b**) NGC 6781; 12″ f/5; 45 min exposure. *Photo: Chris Schur*

Object	NGC 6804
Other names	H VI 38
Type	PLNNB
Mag	12.4
Size	63″×50″
Class	4(2)
Central star	14.2
Surface brightness	11.0
Constellation	Aql
RA	19 31.6
Dec	+09 13
Tirion	16
U2000	207
Description	cB, S, iR, rrr

In the Herschel catalog this planetary is misclassified as open cluster, which must be a misprint. This has been a favorite of mine because it changes its appearance depending on the night, aperture, filters, observer or whatever. Try it for yourself.

6 in.	f/6	S=7 T=8	Excellent site

25×—Extremely faint, pretty small, not brighter in the middle, round.
65×—At this power it can be held steady, but it is still a difficult object in the RFT. In moments of good seeing the "comet" shape of this object can be glimpsed. Most of the time it is round, however.
100×—This is the best view, providing the most contrast against the night sky. And this is on a very good night. Adding the UHC filter does not help much; there are few photons to filter.

13 in.	f/5.6	S=6 T=7	Excellent site

100×—Easy to spot, pretty bright, pretty small, round at this power.
330×—There are two stars involved within the nebulosity, a faint one in the center, a pretty bright one at the eastern edge. This object needs high power for maximum detail. Averted vision elongates the nebula, and the pretty bright star involved also makes the nebula appear comet-shaped, yet with direct vision it is round. Adding the UHC filter makes the stars almost disappear, but the nebulosity is much more contrasty and easy to see.

36 in.	f/5	S=7 T=9 Mc Donald Obs.	Excellent site

300×—There are seven stars involved within the nebula, three of them very faint and only held for 20 % of the time. The central bright region is diamond-shaped. This bright area is mottled and shows lots of fine detail, bright and dark markings that are small, but held steady. The bright central area is surrounded by a fainter region of nebulosity which makes the entire object round. Adding the UHC filter shows this nebula as round all the time and adds to the contrast between light and dark areas. A unique and fascinating object (See Fig. 12.14).

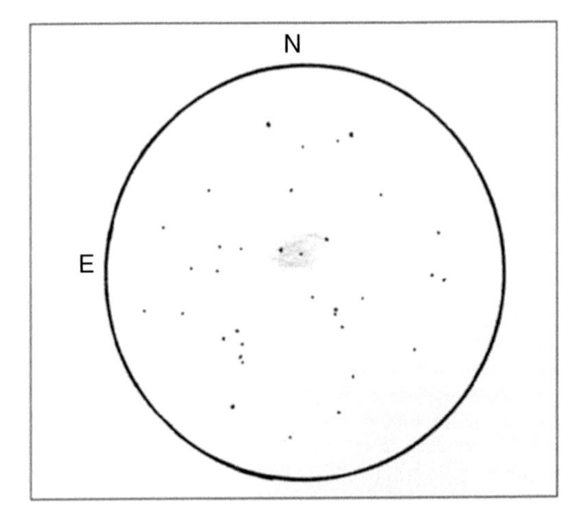

Fig. 12.14 13" f/5.6; 330×; UHC filter

Object	NGC 7009
Other names	PK37-34.1
Type	PLNNB
Mag	8.3
Size	28″ × 12″
Class	4(6)
Central star	12.8
Surface brightness	6.2
Constellation	Aqr
RA	21 04.2
Dec	−11 22
Tirion	16
U2000	300
Description	!!! vB, S, elliptic

NGC 7009 is the Saturn Nebula, a famous planetary with outer ansae (wing-like projections) and a bright inner disk. It was discovered by William Herschel in 1782 but Lord Rosse was the first to see the extending ansae. The projections reminded him of the planet Saturn and he gave this object its nickname. Amateurs have been trying to duplicate that observation ever since (See Fig. 12.15).

6 in. f/6 S=7 T=8 Excellent site

60×—Pretty bright, Pretty small, little elongated, easily seen as non-stellar, a tiny gray dot.

100×—A hint of the greenish color is seen, but it is subtle with the small scope. Even on a night this good, I cannot see the central star or the ansae.

13 in. f/5.6 S=6 T=7 Excellent site

220×—Bright, pretty large, elongated 1.8×1 in PA 75, nice aqua color is easy, averted vision makes the nebulosity about 5× larger.

330×—Center section is quite dark, but no central star is seen.

The ansae are seen about 30 % of time; averted vision helps.

440×—Central star now seen about 20 % of the time. It flashes into view and then is gone for a while.

16 in. f/8 S=7 T=7 Good site
Richard and Helen Line's Obs.

150×—The nebula is bright, pretty small, somewhat elongated and light green.

225×—Shows the ansae as faint projections from the bright central section.

400×—The central star is obvious and the ansae stand out more clearly. One of the bright spots along the ansae (Helen Lines calls them "wing tanks") is visible at this higher power. UHC filter did not help with either the central disk or the ansae detail. Several observers, myself included, saw the nebula as light green, without the UHC filter installed.

36 in.	f/5	S=6 T=8	McDonald observatory	Excellent site

300×—My best view ever, a florescent, aqua-colored oval with a blazing central star. The ansae or wings of the Saturn nebula are immediately obvious. The eastern extension of the ansae has a bright knot or "wing tank" that is easy and obvious at this power.

The western "wing tank" is fainter, so it is seen better with averted vision; both are seen as thickenings within the ansae. The central section is elongated 2.5×1 and there is a small dark region which surrounds the central star. WOW!

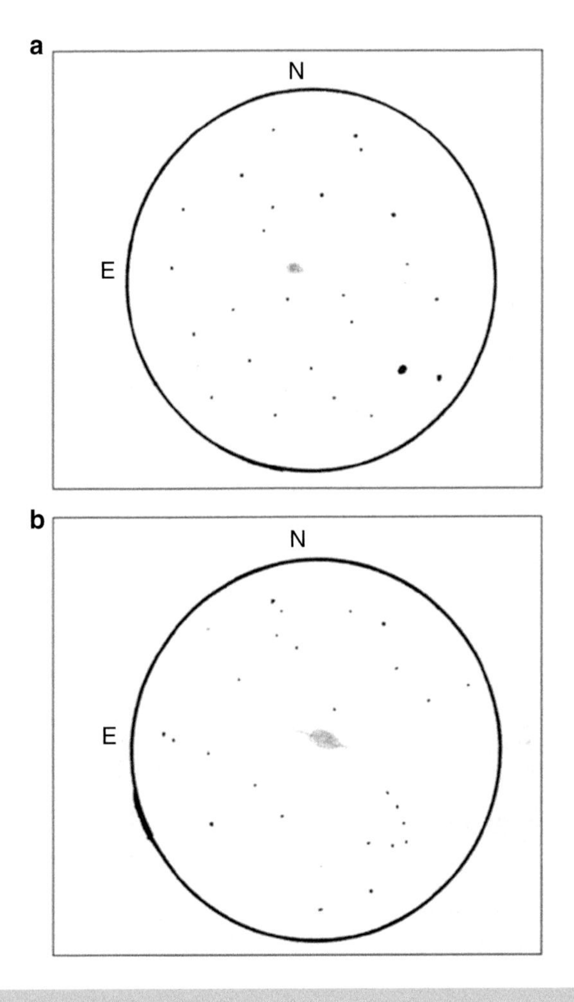

Fig. 12.15 (**a**) 6″ f/6; NGC 7009; FOV 45′; MAG 100×. (**b**) 13″ f/5; NGC 7009; FOV 20′; MAG 220×. (**c**) 36″ f/5; NGC 7009; FOV 12′; MAG 375×. (**d**) NGC 7009; 12″ Lx 200 at f/10; ST 7E CCD. *Photo: Larry E. Robinson*

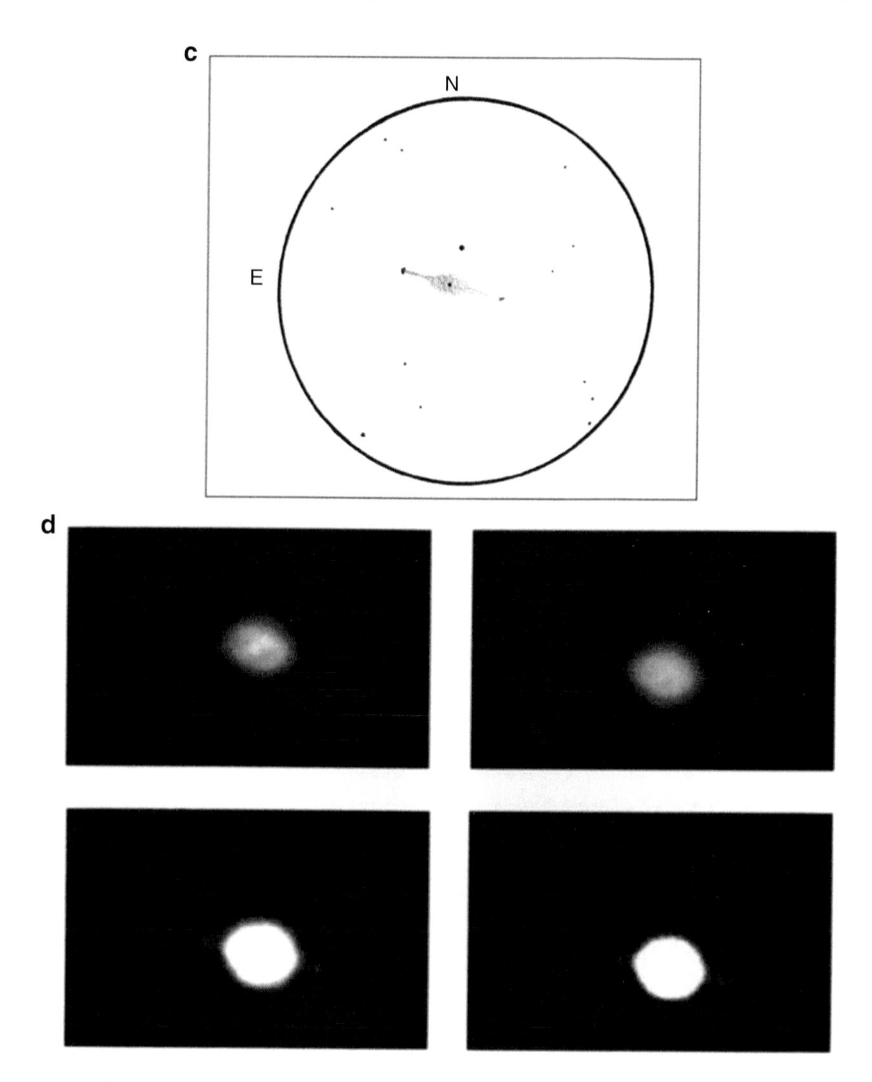

Fig. 12.15 (continued)

Chapter 13

What Can Be Seen in an Open Star Cluster?

So far, many of the objects we have been discussing are created at the end of a star's life. Let's move to the other end of the scale and observe some open star clusters, the places where stars are born. The dust and gas of an emission nebula is pulled toward its center by the relentless force of gravity. Over long periods of time, condensations form within these nebulae and now gravity really has something to pull together. The elements in the nebula are finally forced so close together that these atoms start fusion reactions, and with that a star is formed (See Fig. 13.1).

S.R. Coe, *Deep Sky Observing*, The Patrick Moore Practical Astronomy Series,
DOI 10.1007/978-3-319-22530-2_13

Fig. 13.1 Alpha Perseus Association, 135 mm f/3.5 lens; 7 min exposure

Several things can be said about open star clusters. First, because they are formed from the same gigantic cloud, they have the same chemical nature. The types and abundances of the elements are the same for all cluster members. Second, because there is much more gas and dust within the disk of the Milky Way, these open clusters are generally seen along the width of Our Galaxy.

During the 1930s R.J. Trumpler was studying the various open clusters while at Lick Observatory, on Mount Hamilton, near San Jose, California. He created a classification scheme that proved to be useful and the modern data for many open clusters included the "Trumpler type". The information below includes the three factors: concentration, range of magnitudes and richness of stars. Even before you observe a cluster in your telescope, you can get an idea of what the cluster should look like from this rating system. It also will allow you to make certain that the cluster for which you are searching is the one you have in the eyepiece.

The reason I have always appreciated a beautiful open cluster comes from spending some time looking for star groupings and the clusters' setting within the Milky Way. Each of these star groupings is unique. Look at the cluster and its surroundings the next time you put one of these lovely gathering places on your observing list (Table 13.1).

Table 13.1 Trumpler classes for open clusters

Concentration	
I	Detached, strong concentration toward the center
II	Detached, weak concentration toward the center
III	Detached, no concentration toward the center
IV	Not well detached from surrounding star field
Range in brightness	
1	Small range
2	Moderate range
3	Large range
Richness	
p	Poor (<50 stars)
m	Moderately rich (50–100 stars)
r	Rich (>100 stars)

An n following the Trumpler type denotes nebulosity in cluster.

Object	**NGC 457**
Other names	**H VII 42**
Type	**OPNCL**
Mag	**6.4**
Size	**13.0′**
Class	**I 3 r**
Number of stars	**80**
Brightest	**08.6**
Constellation	**Cas**
RA	**01 19.1**
Dec	**+58 20**
Tirion	**1**
U2000	**36**
Description	**Cl, B, L, pRi, *7,8,10**

Even though this bright cluster surrounds the star Phi Cas, that bright star is not a member of the cluster. The shape of this grouping looks much like the dolls made by Navajo tribe in the US Southwest. Therefore in the Saguaro Astronomy Club we have always called this the "Kachina Doll" cluster.

11×80 binoculars	S=7 T=7	Excellent site

11×—Bright, pretty large, 10 stars counted, including Phi Cas. The "Kachina Doll" outline is easy to see, even at this low power.

6 in	f/6	S=7 T=9	Excellent site

100×—Bright, large, not compressed, pretty rich, several orange stars involved, two bright stars and cluster; Phi Cas is light yellow. 48 total stars counted.

13 in.	f/5.6	S=7 T=8	Excellent site

100×—Bright, large, pretty rich, compressed. 63 stars counted including Phi Cas, a light-yellow star at the edge of the cluster. There is another bright star near Phi that gives the effect of having two glowing eyes looking back at the observer. The Kachina Doll shape is easily seen, two sparking eyes and the rest of the cluster outlines outstretched arms with feathers.

150×—The cluster is about 3/4 of the field and it appears that all the stars are resolved. I counted 78 stars, but there is no fuzzy background. There is a nice, orange star on the northwest side of the cluster.

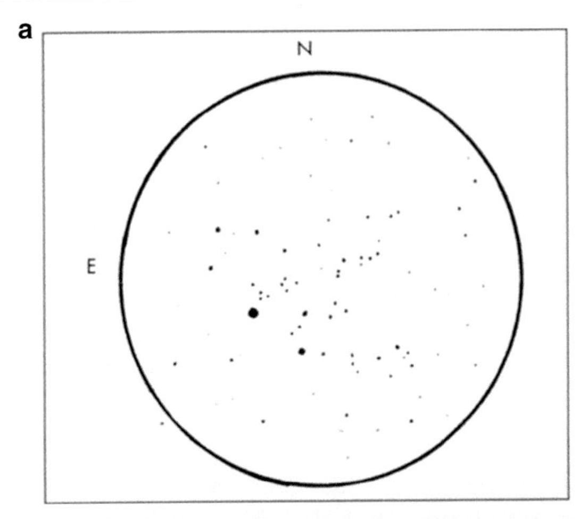

Fig. 13.2 (**a**) 13″ f/5.6; NGC 457; FOV 30′; MAG 100×. (**b**) ED 80 f/7; Canon T2i camera; 4 min exposure

Object	NGC 752
Other names	**H VII 32**
Type	**OPNCL**
Mag	**5.7**
Size	**50.0′**
Class	**III 1 m**
Number of stars	**60**
Brightest	**09.0**
Constellation	**And**
RA	**01 57.8**
Dec	**+37 41**
Tirion	**4**
U2000	**92**
Description	**Cl, vvL, Ri, *L, & S, C**

11×80 binoculars S=7 T=7 Excellent site

11×—Bright, large, not compressed, an obvious cluster. Bracing my elbows against the truck roof, I can count 15 stars resolved in this big cluster.

6 in. f/6 S=5 T=5 Mediocre site

65×—Bright, large, not compressed, somewhat rich, 22 stars counted with several nice chains and a fuzzy background. There is a bright triple star on the south side.

6 in. f/6 S=7 T=7 Excellent site

25×—Bright, very large, rich, compressed. I counted 62 stars resolved, even at this lowest magnification. Many bright, and pretty bright members in this nice cluster. The RFT fits this cluster very nicely. A nice triple star is involved with in the cluster; two components are yellow and a third is light blue in color.

13 in. f/5.6 S=7 T=8 Mediocre site

60×—Very bright, very, very large, round, not compressed, and rich.

I counted 65 stars as members; this huge cluster has stars of 10th to 13th mag. It appears that the larger scope has resolved all the stars which can be seen in this cluster. There is no hazy background of unresolved stars (See Fig. 13.3).

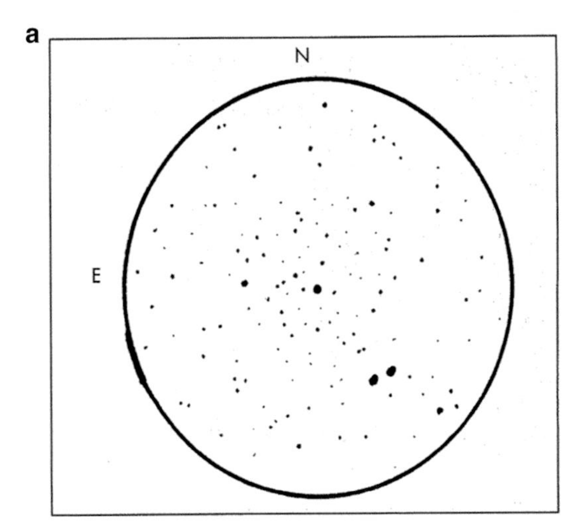

Fig. 13.3 (**a**) 6″ f/6; NGC 752; FOV 1.2°; MAG 40×. (**b**) 300 mm lens f/4.5; Canon Xt camera; 6 min exposure

Object	**NGC 1528**
Other names	**H VII 61**
Type	**OPNCL**
Mag	**6.4**
Size	**24.0′**
Class	**II 2 m**
Number of stars	**40**
Brightest	**08.8**

(continued)

(continued)

Constellation	**Per**
RA	**04 15.4**
Dec	**+51 14**
Tirion	**1**
U2000	**39**
Description	**Cl, B, vRi, cC**
Notes	**80* mag 8…,1° NNW from b1 Per.**

10×50 binoculars	S=7 T=7	Excellent site

10×—Easy to see, no stars resolved, but a pretty bright fuzzy spot in the binoculars on a fine night.

6 in.	f/6	S=7 T=7	Excellent site

25×—Bright, pretty large, rich, considerably compressed, 28 stars counted in the cluster even at this low power with the 6 in. The cluster stands out well from the Winter Milky Way through Perseus.

65×—Stars counted, still quite compressed toward the middle; there are half a dozen nice double stars within this lovely cluster.

13 in.	f/5.6	S=6 T=6	Good site

135×—Bright, large, rich and somewhat compressed cluster. 85 stars counted with many in lovely chains. Several dark lanes wind their way through the cluster. It almost fills the field of view. There are stars of a wide variety of colors in this lovely cluster: blue, yellow, orange, white and gold.

This very nice cluster is a favorite of mine and could have been a Messier object if he had only swept this area (See Fig. 13.4).

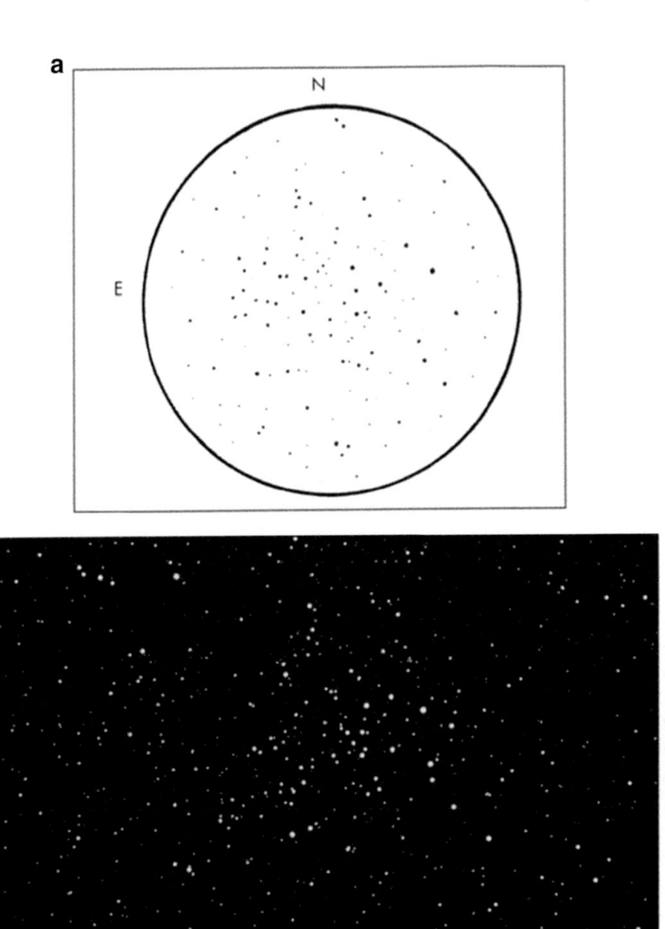

Fig. 13.4 **(a).** 13″ f/5.6; NGC 1528; FOV 30′; MAG 100×. **(b)** C11 at f/6.3; ST 7E CCD. *Photo: Robert Kuberek*

Object	**NGC 1857**
Other names	**H VII 33**
Type	**OPNCL**
Mag	**7.0**
Size	**6′**
Class	**II 2 m**
Number of stars	**40**
Brightest	**07.4**
Constellation	**Aur**
RA	**05 20.2**
Dec	**+39 21**
Tirion	**5**
U2000	**66**
Description	**Cl, pRi, pC, st7…**

6 in. f/6 S=7 T=7 Excellent site

40×—Pretty bright, small, little compressed, not rich. This fuzzball is just seen as a cluster at this power.

65×—Does not help much, seven stars resolved with direct vision, averted vision adds five more, but not a well-resolved cluster with this aperture.

13 in. f/5.6 S=7 T=8 Excellent site

100×—Easily seen as cluster, a pretty bright spot in the Milky Way.

150×—Pretty bright, pretty large, pretty rich, somewhat compressed, 38 stars counted, including an 8th mag yellow star near center of cluster.

There are many faint members that are easier at high power, including several beautiful chains of stars (See Fig. 13.5).

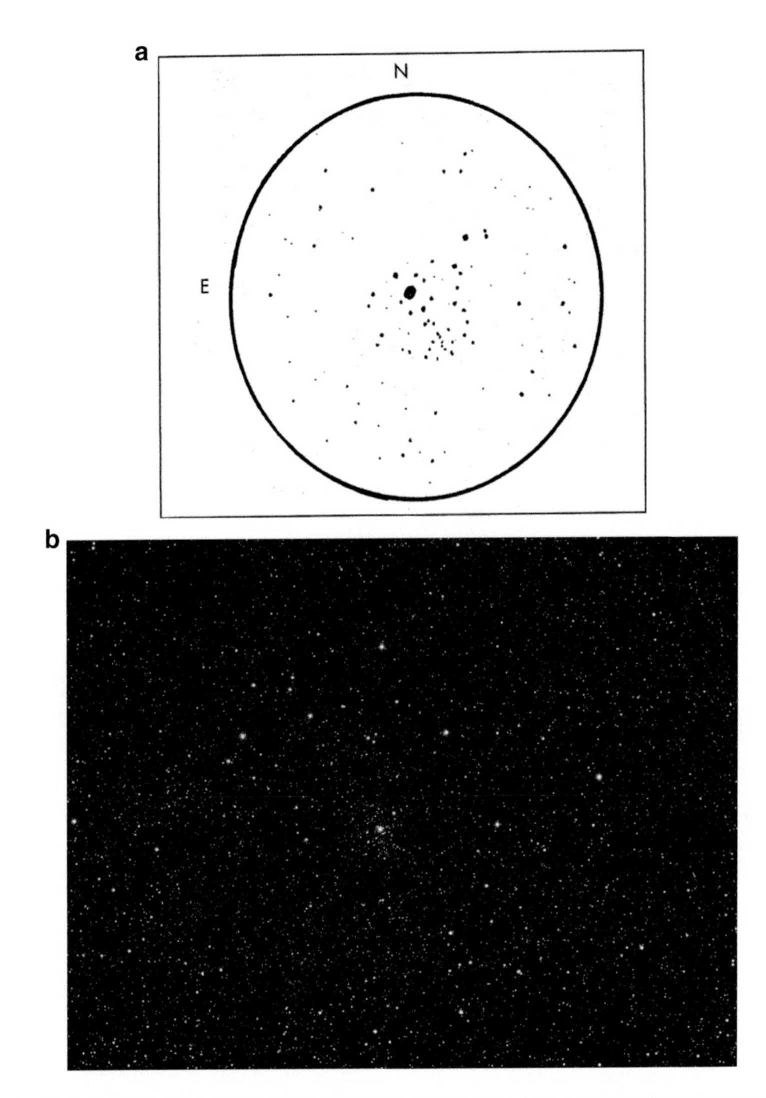

Fig. 13.5 (**a**) 13″ f/5.6; NGC 1857; FOV 20′; MAG 150×. (**b**) AT90EDT SBIG ST8300M; *Photo: Dan Crowson*

Object	**M37**
Other names	**NGC 2099**
Type	**OPNCL**
Mag	**5.6**
Size	**24.0′**
Class	**II 1 r**
Number of stars	**150**
Brightest	**09.2**
Constellation	**Aur**
RA	**05 52.4**
Dec	**+32 33**
Tirion	**5**
U2000	**98**
Description	**!!Cl, Ri, pCM, st L & S**

The total population of this very rich cluster is over 500 stars, making it one of the richest star clusters in the sky.

10×50 binoculars S=7 T=7 Excellent site

10×—Bright, compressed, much brighter middle. There are three stars resolved across the face of a fuzzy area in the winter Milky Way. Easy to see, well detached from the background of stars. Averted vision makes it grow in size and shows about five more stars at the limit of the binoculars

6 in. f/6 S=7 T=7 Excellent site

25×—Bright, large, very rich, compressed, 25 stars are resolved.

40×—42 stars counted, including a lovely light-orange star on the north side.

There are many beautiful chains of stars winding their way through the cluster. A great view.

13 in. f/5.6 S=7 T=7 Excellent site

150×—Very bright, large, very rich, pretty compressed. 146 stars counted, many beautiful chains of stars and dark lanes through the cluster, dark-orange star of about 10th mag on north side, in a dark area.

220×—Many faint stars now resolved, lots of pairs and chains, all broken up by dark lanes. Even though the cluster takes up about 3/4 of the field of view, it is still seen as a star cluster with more stars than the surrounding Milky Way (See Fig. 13.6).

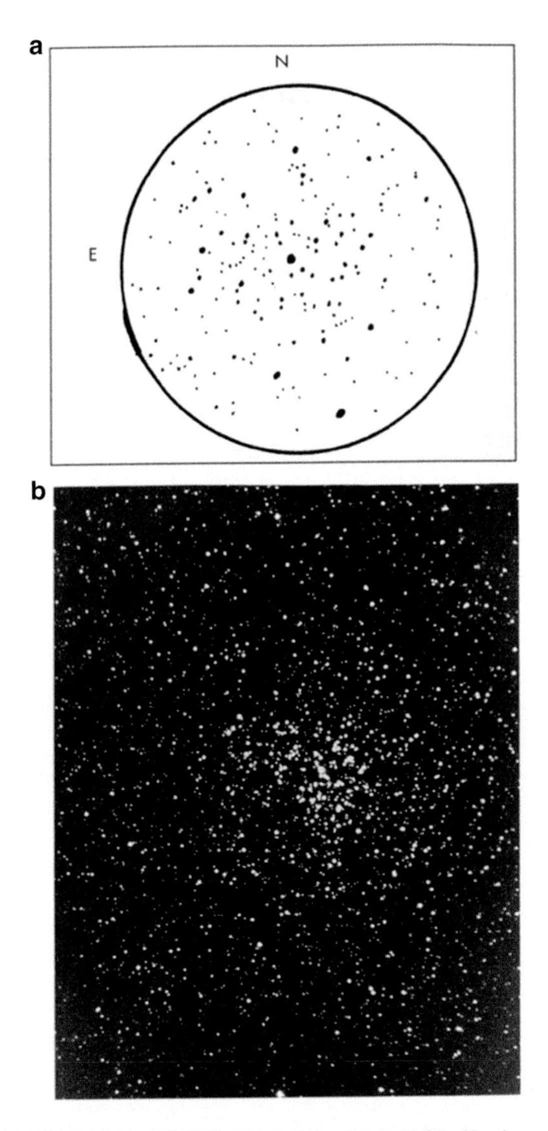

Fig. 13.6 (**a**) 13″ f/5.6; M 37; FOV 30′; MAG 100×. (**b**) C 14 f/7; 40 min exposure. *Photo: David Healy*

Object	**NGC 2301**
Other names	**H VI 27**
Type	**OPNCL**
Mag	**6.0**
Size	**12.0′**
Class	**I 3 m**
Number of stars	**80**
Brightest	**08.0**
Constellation	**Mon**
RA	**06 51.8**
Dec	**+00 28**
Tirion	**11**
U2000	**228**
Description	**Cl, Ri, L, iF, st L & S**

Many of these beautiful sites in the Milky Way have a special treat and this cluster is one of those. It is a pretty nice star grouping, but it also is blessed with a surprise. Right in the middle of this open cluster is a lovely blue-and-gold double star. If you have never made room on your observing list for NGC 2301, give it a try. This is a personal favorite.

6 in.	f/6	S=7 T=7	Excellent site

165×—Pretty bright, pretty large, pretty compressed and pretty rich. 27 stars counted. The yellow-and-blue pair is not as obvious in the RFT. In a very rich Winter Milky Way field of view.

13 in.	f/5.6	S=7 T=8	Excellent site

100×—Bright, large and pretty rich with 40 members counted. It is easy to pick out in the finderscope. The aspect of this cluster which makes me return each winter is a lovely blue-and-gold double star right in the center. There is a clear area around the double star (See Fig. 13.7).

(continued)

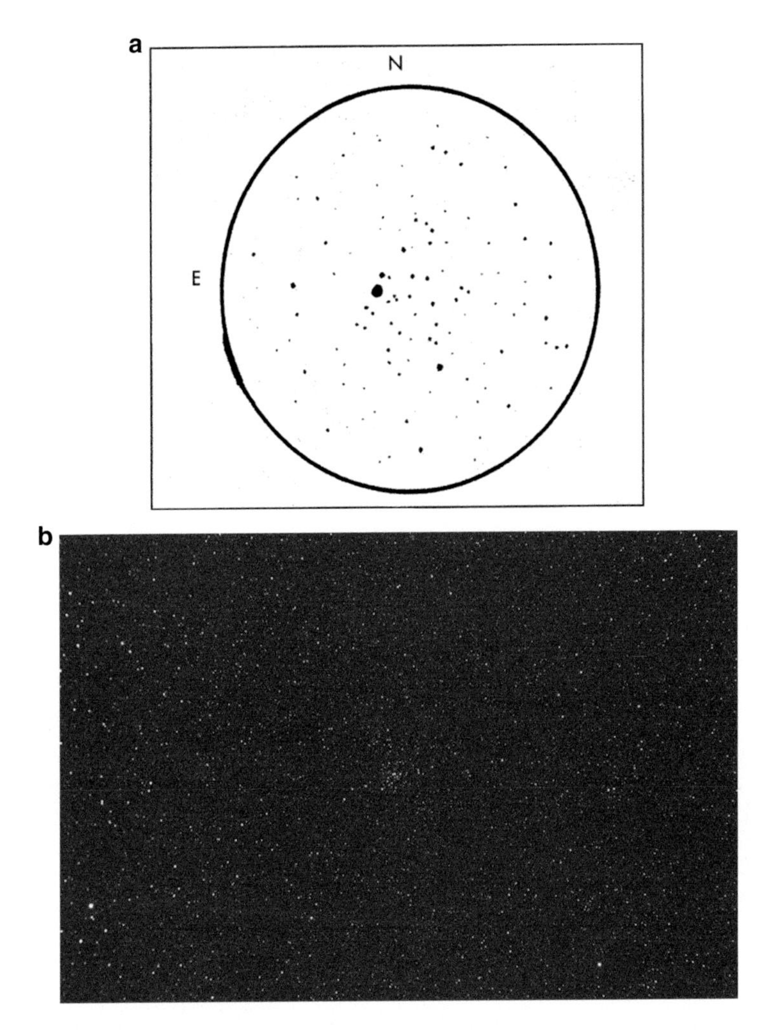

Fig. 13.7 (**a**) 13″ f/5.6; NGC 2301; FOV 20′; MAG 150×. (**b**) ED 80 f/6; Canon 2Ti camera; 4 min exposure

Object	NGC 2355
Other names	H VI 6
Type	OPNCL
Mag	9.7
Size	9.0′
Class	II 2 p
Number of stars	40
Brightest	13.0
Constellation	Gem
RA	07 16.9
Dec	+13 47
Tirion	12
U2000	184
Description	Cl, pS, pRi, mC,* 15..16

10×50 binoculars S=8 T=9 Excellent site

110×—A small and unresolved soft glow, not well detached, next to a pretty bright star.

6 in. f/6 S=6 T=6 Good site

65×—Pretty bright, pretty small, much compressed, fan shaped cluster. 10 stars resolved of magnitude 12 and fainter with direct vision. Averted vision shows another 10 stars that are extremely faint with this aperture. There is a very faint background glow of unresolved stars. Higher powers did not show any more detail on a mediocre night.

13 in. f/5.6 S=8 T=9 Excellent site

100×—Pretty bright, pretty small, pretty rich, considerably compressed. 26 stars counted with fuzzy background. Averted vision shows the background as grainy, so use higher power.

150×—32 stars counted, a much better view. The cluster is now about 50 % of the field of view. Double star on the east side is now seen as triple with delicate 13th mag companion. The "double" star is yellow and light blue, the faint one white. All the stars seem resolved. 220× adds nothing (See Fig. 13.8).

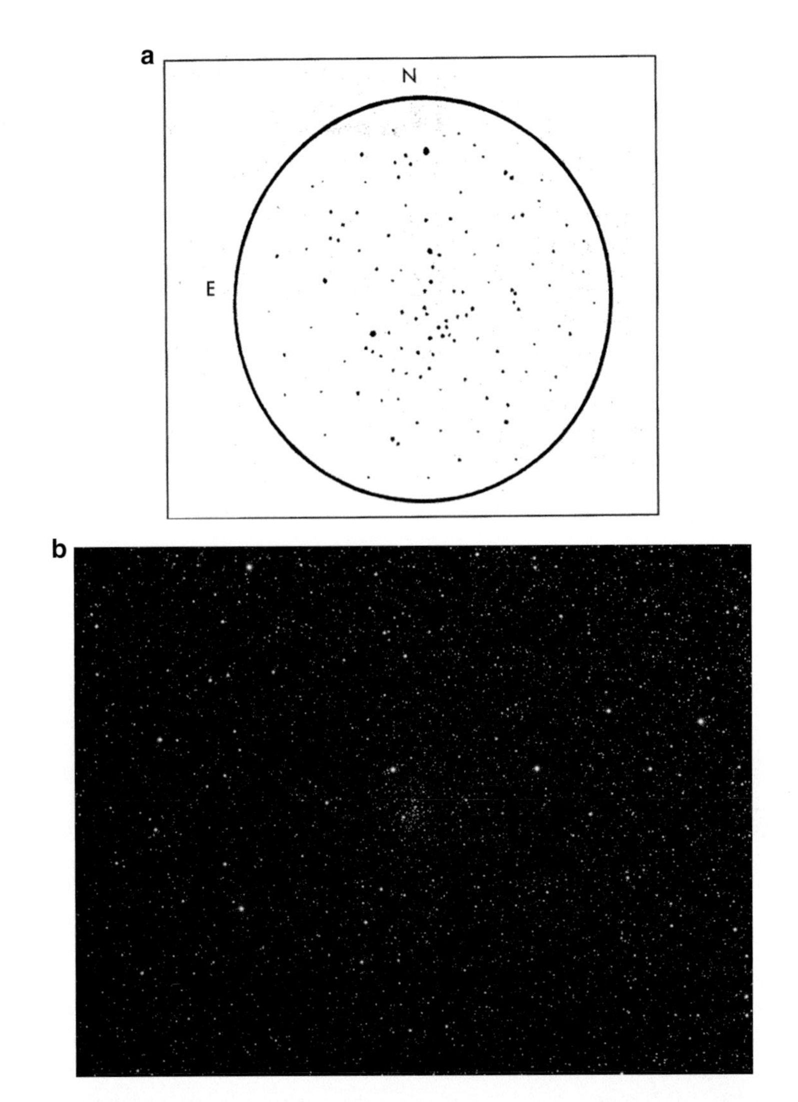

Fig. 13.8 (**a**) 13″ f/5.6; NGC 2355; FOV 20′; MAG 150×. (**b**) AT90EDT SBIG ST8300M; *Photo: Dan Crowson*

Object	**NGC 2362**
Other names	**H VII 17**
Type	**OPNCL**
Mag	**4.1**
Size	**8.0′**
Class	**I 3 p n**
Number of stars	**60**
Brightest	**04.4**
Constellation	**CMa**
RA	**07 18.8**
Dec	**−24 57**
Tirion	**19**
U2000	**319**
Description	**Cl, pL, Ri, (30 CMa)**

If you are having trouble finding deep-sky objects, this one is easy. This cluster includes Tau CMa, a naked-eye star on even a mediocre night.

6 in.	f/6	S=7 T=7	Excellent site

40×—Tau CMa is a single star and there are 19 companion members in the cluster

100×—Pretty large, pretty bright, pretty rich, considerably compressed, stars mags 10…. Tau is double, the companion star is faint and light blue, located to the east of Tau, the primary star.

17.5 in.	f/4.5	S=6 T=7	Excellent site

165×—Pretty bright, pretty large, somewhat compressed, round, consists of Tau CMa and about 45 stars. Tau has dark band around it, then the cluster members fan out from this dark ring. Tau has two companions that form almost a straight line. Tau is white, its two companions bluish and to one side. Having a bright triple star in the center of a cluster is unique and I return to this object often (See Fig. 13.9).

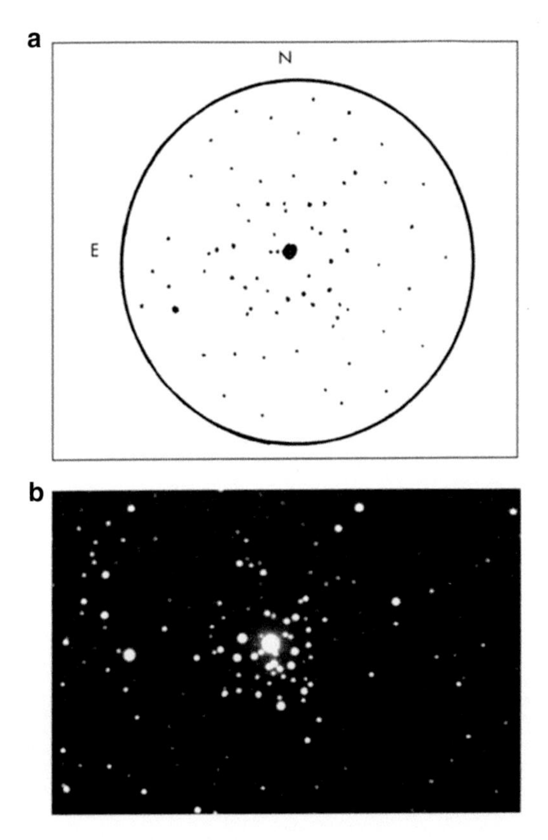

Fig. 13.9 (**a**) 13″ f/5.6; NGC 2362; FOV 25′; MAG 150×. (**b**) 12″ Lx 200 at f/7; ST 7E CDD; 6 min exposure. *Photo: Larry E. Robinson*

Object	NGC 2477
Type	**OPNCL**
Mag	**5.8**
Size	**27.0′**
Class	**I 2 r b**
Number of stars	**200**
Brightest	**12.0**
Constellation	**PUP**
RA	**07 52.3**
Dec	**−38 33**
Tirion	**19**
U2000	**362**
Description	**!, Cl, B, Ri, L, IC,*12**

Harlow Shapley worked on this very rich cluster and determined that there are 300 stars down to 12th mag. This is an extraordinary value, making NGC 2477 one of the most star-rich galactic clusters in the sky.

11 × 80 finderscope S = 6 T = 7 Excellent site

11×—Bright, very large, round, suddenly brighter middle. 12 stars can be resolved, even in the large finderscope!

6 in. f/6 S = 7 T = 8 Good site

40×—19 Stars counted, very bright, very large, very rich, pretty compressed, stars 10 m…. Some dark lanes on the south side and beautiful chain of stars on the northeast side.

65×—Best view for resolving stars, but better as cluster at lower power. 28 stars counted, nice matched double star about 11th mag on north side. Dark lanes are more prominent at higher power.

13 in. f/5.6 S = 7 T = 8 Excellent site

60×—Very bright, very large, compressed, 60 stars resolved. There are two round dark markings in the cluster on the south side. A nice wide double star is on the north edge.

100×—Very, very rich, very compressed. 180 total members are estimated by counting 45 stars in the northeast quadrant of this cluster. There are many lovely dark lanes that wind their way through several star chains and an ever-present glowing background of more stars. There are two round dark markings in the cluster on the south side and a nice, wide double star on the north edge. A great view, still an obvious cluster with *lots* of members. Both 150× and 220× don't show fainter stars, but the close pairs of double stars in the cluster are being split and at 220× the cluster overfills the field of view in the scope (See Fig. 13.10).

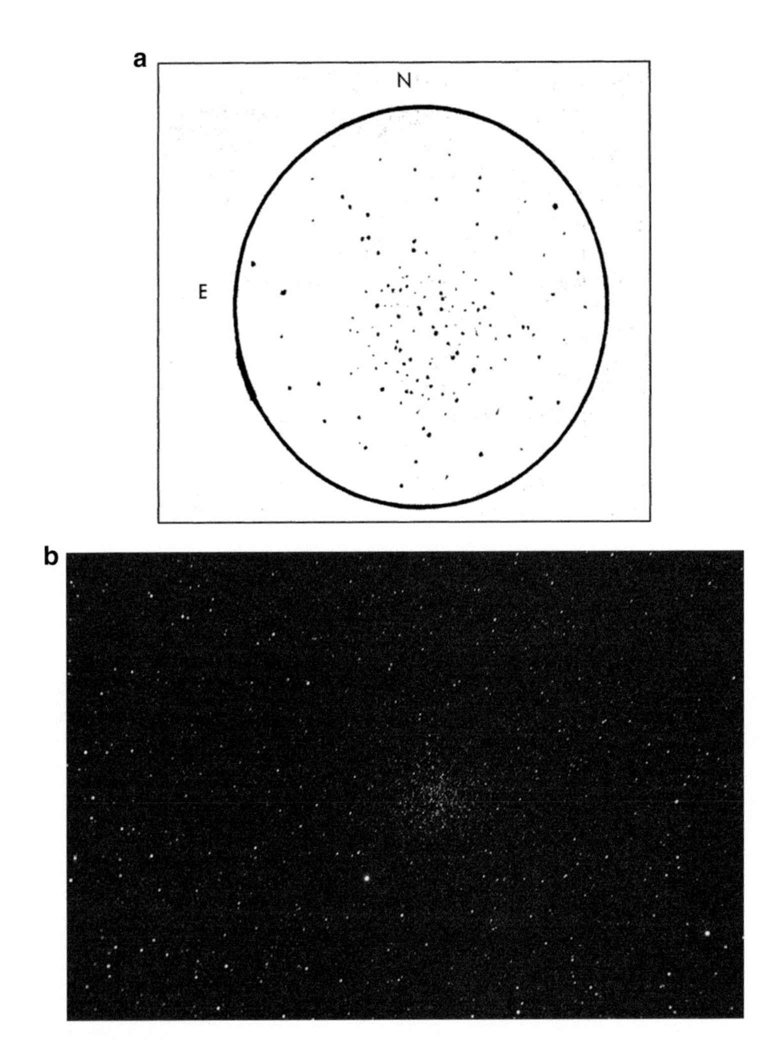

Fig. 13.10 (**a**) 6″ f/6; NGC 2477; FOV 50′; MAG 65×. (**b**) ED 80 f/7; Canon 2Ti; 4 min exposure

Object	NGC 6451
Other names	H VI 13
Type	OPNCL
Mag	8.2
Size	8.0′
Class	II 1 p n
Number of stars	80
Brightest	12.0
Constellation	Sco
RA	17 50.7
Dec	–30 13
Tirion	22
U2000	377
Description	Cl, pL, pRi, bifid, st12

Some clusters are not noteworthy because of an abundance of stars or nebulosity, but how the stars are arranged provides a fascinating and fun view at the telescope. NGC 6451 is an otherwise unremarkable cluster of stars, except for the fact that a wide dark lane run s down over half the cluster. I have heard it called the "Jack Horner cluster" because it does look like he stuck his in thumb and pulled out a plum—well, at least he pulled out the stars that should fill in the center of the cluster.

10×50 binoculars S=6 T=6 Good site.

10×—Just seen as a fuzzy spot

6 in. f/6 S=6 T=6 Good site

40×—Stars in front of a fuzzy background, elongated 1.5×1 in PA 0. A pretty obvious cluster, even at low power.

100×—Pretty bright, pretty large, pretty rich, somewhat compressed. 14 stars of magnitudes 11 and fainter are counted within the cluster. The fuzzy background of unresolved stars is still prominent. There is a dark lane down the middle and some of the stars line the boundary of this dark feature.

17.5 in. f/4.5 S=6 T=7 Good site

60×—Seen as a fuzzy spot at low power, 12 stars resolved.

165×—Bright, pretty large, pretty rich, compressed open cluster. It includes a close triple star which appears nebulous at low powers and is resolved at 320×. Fifty members were counted in the cluster and it includes a dark lane almost down its middle. A dozen or so of the faint members of the cluster outline the dark lane, so it appears even more prominent than otherwise, a fascinating feature (See Fig. 13.11).

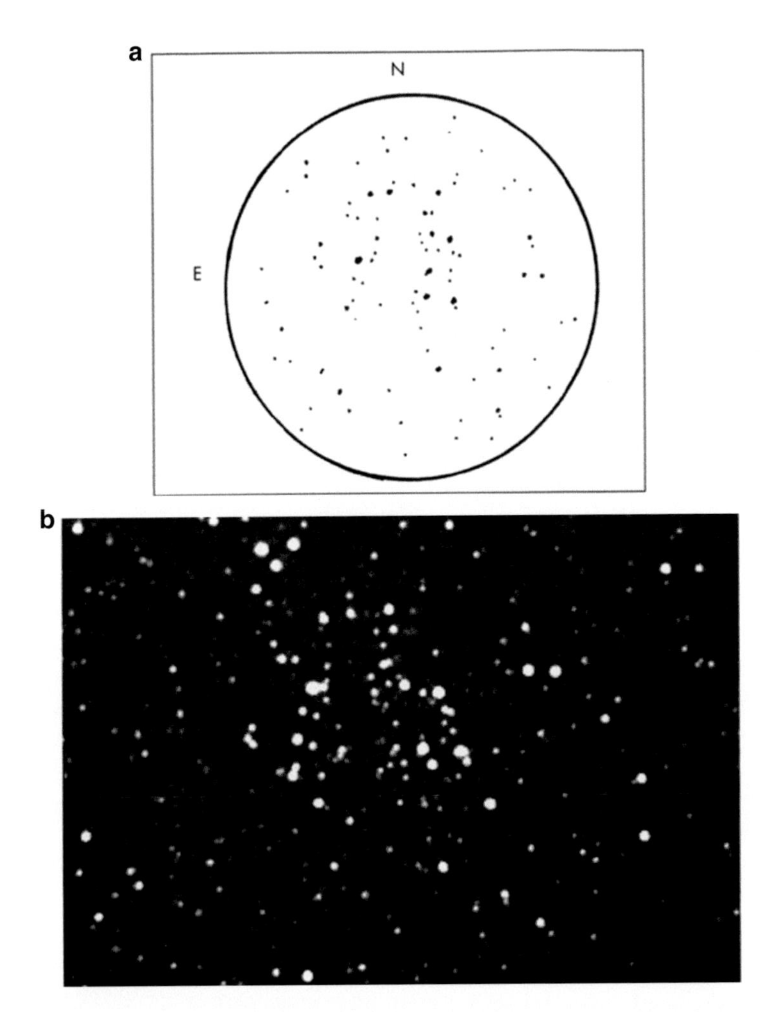

Fig. 13.11 (**a**) 13″ f/5.6; NGC 6451; FOV 20′; MAG 220×. (**b**) 12″ L× 200 at f/7; ST 7E CCD; 6 min exposure. *Photo: Larry E. Robinson*

Object	NGC 6469
Other names	
Type	**OPNCL**
Mag	**8.2**
Size	**12.0′**
Class	**III 2 p**
Number of stars	**50**
Constellation	**Sgr**
RA	**17 52.9**
Dec	**−22 21**
Tirion	**22**
U2000	**339**
Description	**Cl, pRi, in Milky Way**

I included this rather ordinary cluster to show that driving to a dark site and being fortunate enough to have it turn out to be a superior night really makes a difference, even in an object that is never going to be a showpiece. Going to all the effort to pack up the scope and drive for two hours can really be worth it, if the weather cooperates.

6 in.	f/6	S=6 T=6	Good site

40×—Cluster, pretty faint, little compressed. 10 stars are resolved, 4 of them very faint. The cluster is seen imposed on a fuzzy background of unresolved stars.
100×—Now 14 stars are resolved, including three nice pairs of stars.

13 in.	f/5.6	S=6 T=7	Good site

150×—Pretty faint, pretty large, not rich, not compressed, 18 stars counted with fuzzy background haze, stars mag 9…12.

13 in.	f/5.6	S=8 T=10	Superior site

100×—Bright, pretty large, pretty rich and somewhat compressed, 29 stars counted. It can just be seen in the 11×80 finder.
150×—Now the cluster fills the central half of the field and 44 stars are resolved, included a dozen at the limit of the 13 in. on a fabulous night. Several double stars are resolved, including one right at the middle of the cluster (See Fig. 13.12).

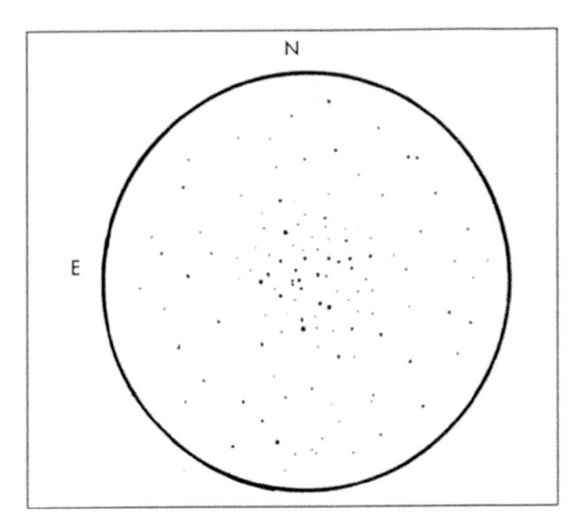

Fig. 13.12 13″ f/5.6; NGC 6469; FOV 30′; MAG 100×

Object	**NGC 6791**
Type	**OPNCL**
Mag	**9.5**
Size	**16.0′**
Class	**II 3 r**
Number of stars	**300**
Brightest	**13.0**
Constellation	**Lyr**
RA	**19 20.7**
Dec	**+37 51**
Tirion	**8**
U2000	**118**
Description	**vF, L, vRi,*F**

6 in. f/6 S=7 T=7 Excellent site

40×—Faint, pretty large, not brighter in the middle and has three stars resolved.

100×—The 8.8 mm eyepiece will resolve six stars and makes this fuzzy cluster larger and more obvious. Clusters like this one, with lots of faint and very faint stars, are never going to be a showpiece with a small telescope.

13 in. f/5.6 S=7 T=8 Excellent site

100×—Pretty bright, pretty large, somewhat compressed, little elongated 1.5×1 in PA 90, 12 stars counted with mottled or "oatmeal" glow in background. Light-yellow star "U Lyr" on northern edge.

150×—Cluster aspect almost gone, tough to see edges of cluster, counted 38 stars as members, 5 extremely faint; averted vision helps cluster stand out. Pretty rich, with lots of dim members (See Fig. 13.13).

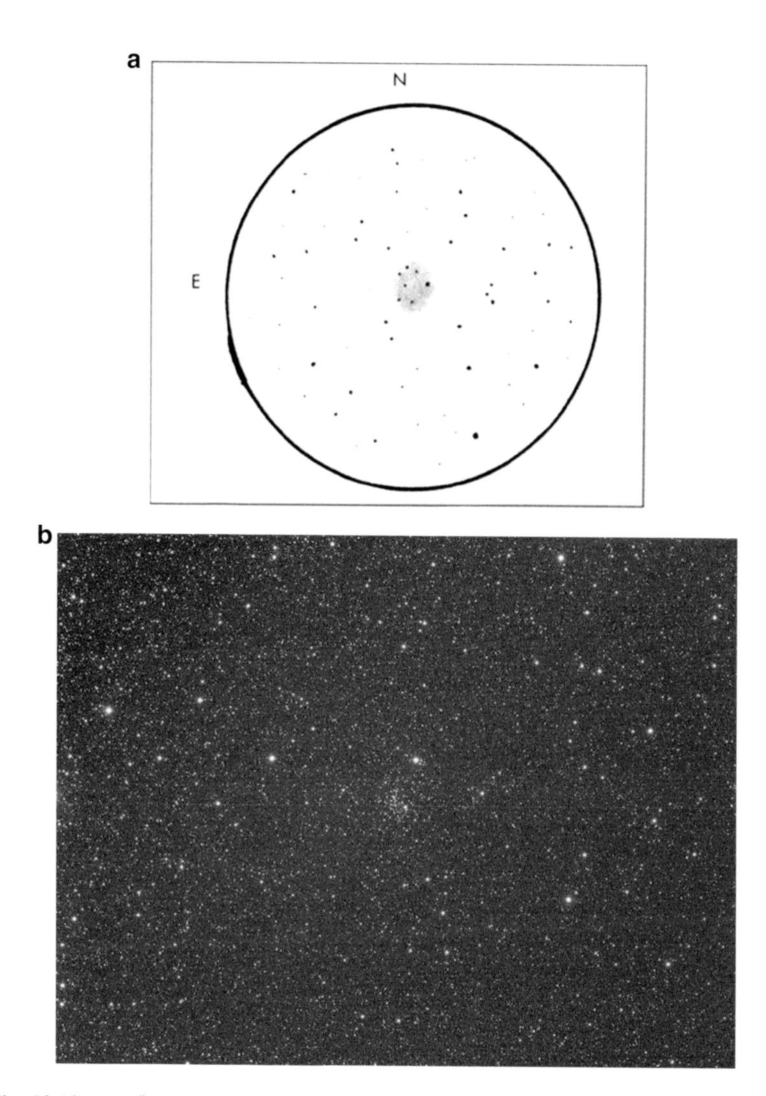

Fig. 13.13 (**a**) 6″ f/6; NGC 6791; FOV 45′; MAG 100×. (**b**) AT90EDT SBIG ST8300M; *Photo: Dan Crowson*

Object	**M 24**
Other names	**IC 4715**
Type	**OPNCL**
Mag	**4.0**
Size	**80′**

(continued)

(continued)

Class	
Constellation	**Sgr**
RA	**18 17.0**
Dec	**−18 35**
Tirion	**22**
U2000	**339**
Description	**vvB, eL, mE, many* of diff mags**

The Small Sagittarius Star Cloud has always fascinated me. This elongated glow in the summer Milky Way is composed of so many stars in such a wide variety of brightness that virtually any optical aid will show a new aspect of this object. The two prominent dark markings at its northern edge are covered in the dark Nebula chapter. The compressed open cluster, NGC 6603, was mistaken for M24 for years; the error was corrected with further scrutiny of Messier's notes.

Naked eye S=6 T=8 Excellent site

1×—An obvious glow, elongated 2.5×1 E–W, somewhat brighter in the middle and wider on the west side. Averted vision makes it grow.

8×25 binoculars S=6 T=8 Excellent site

8×—Seventeen stars involved. The dark markings on the north side are just seen, looking like dark, inhuman eyes. The star cloud is nicely framed by the bright and dark swath of the Milky Way. With their very wide field, the small binoculars provide a fine view, (7°) framing this big object nicely.

10×50 binoculars S=6 T=8 Excellent site

10×—56 Stars resolved, including several curved chains of stars.

This silvery, elongated star cloud is beautiful. Several dark lanes in the Milky Way are nearby. The cluster NGC 6603 stands out quite well.

11×80 binoculars S=7 T=7 Excellent site

11×—This huge star cloud is the central 70 % of the field; 52 stars are resolved.

The dark lanes are easy on both sides of M24; they "frame" the star cloud.

Several yellow and one light-orange star are seen with direct vision. There is a silvery background of unresolved stars

6 in.	f/6	S=6 T=8	Excellent site

25×—Very, very bright, extremely large, somewhat brighter in the middle, elongated, very, very rich. Even at the lowest power available to me in the RFT, this magnificent star cloud overflows the field of view! The glowing, silvery face of M24 is criss-crossed by beautiful curved chains of stars. Many are bright enough to show color, even in the 6 in. The four prominent colors are white, blue-white, yellow and orange.

40×—Extremely bright, extremely large, extremely rich, a huge oval of stars, obviously naked-eye. NGC 6603 is involved and at this power this cluster is just a fuzzy spot with a bright, compressed core. There is a dark-orange star at the edge of the cluster. The entire star cloud is over two fields of view in the RFT, making it at least 3° in total size. 56 stars are counted in the central 10 % of the star cloud, so there are about 500 stars resolved across the face of this unique object. Many lovely, delicate chains of stars criss-cross the body of the Star Cloud, like sparkling diamonds on black velvet. Overwhelming numbers of stars in curving chains with dark lanes cutting through them. A fascinating interplay of stars and dark lanes on the southwest end of the star cloud (See Fig. 13.14).

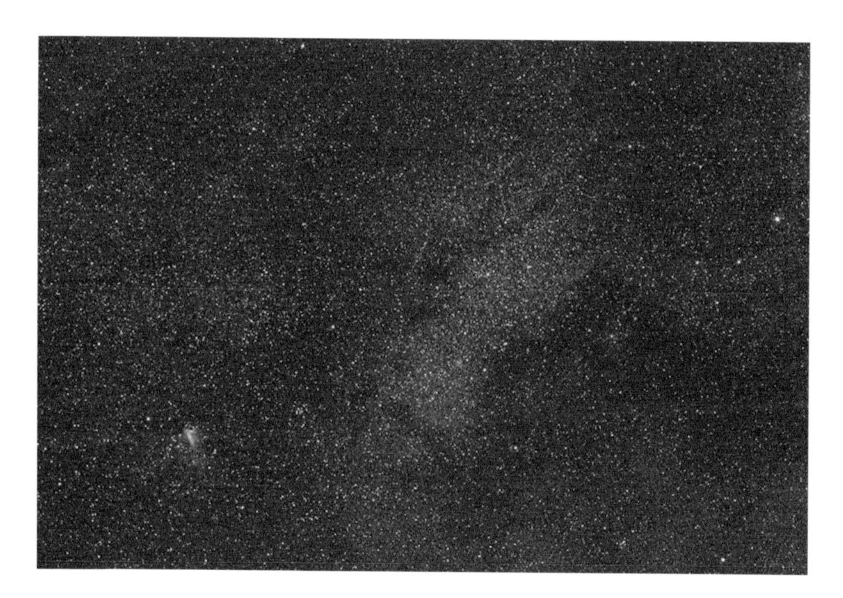

Fig. 13.14 M24 200 mm f/4.5; Canon Xt camera; 5 min exposure

Object	NGC 7686
Other names	H VIII 69
Type	OPNCL
Mag	5.6
Size	15.0′
Class	IV 1 p
Number of stars	20
Brightest	06.2
Constellation	And
RA	23 30.2
Dec	+49 08
Tirion	9
U2000	88
Description	Cl, p, lC, st 7…11

This cluster is included to demonstrate that you should not give up on an object just because you have observed it before and didn't see much. Often a better night, or more aperture, or both, will yield some excellent detail that was missed before. This one is certainly not a showpiece, but it is worthy of some observing time.

6 in. f/6 S=6 T=6 Good site

40×—Just seen as four stars near on 8th mag star.

65×—Pretty faint, small, not compressed. Eight stars seen with direct vision; averted vision odds four more very faint members. Not much of a cluster aspect.

13 in. f/5.6 S=6 T=6 Good site

100×—Pretty bright, pretty compressed, round, surrounds a yellow 8th mag star, reminds me of Tau CMa and NGC 2362. 21 members counted, so this cluster is not rich.

13 in. f/5.6 S=7 T=8 Excellent site

100×—Pretty bright, large, pretty compressed, 37 members counted. One pretty bright member is a very nice orange color. The beautiful chains of stars form a spiral pattern outward from the center. The stars are of magnitudes 8 through 12. The Milky Way does not have a very high star density here, so it is well detached.

150×—Brings out five more very faint members, but that seems to be all there is to see. Also brings out a nice star chain and another triple star near the bright central star. More power does not help (See Fig. 13.15).

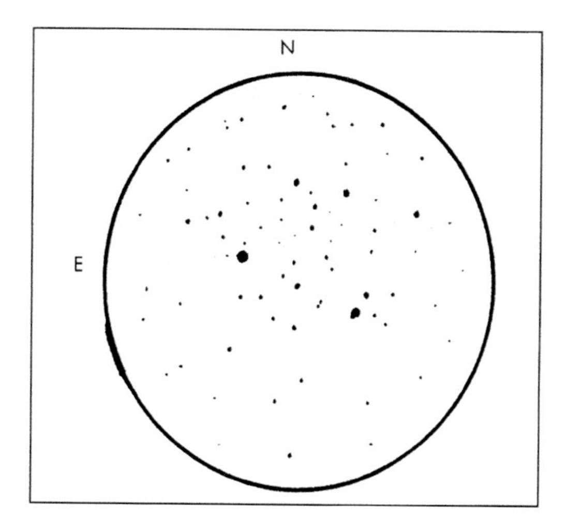

Fig. 13.15 13″ f/5.6; NGC 7686; FOV 20′; MAG 150×

Object	**NGC 7789**
Other names	**H VI 30**
Type	**OPNCL**
Mag	**6.7**
Size	**16.0′**
Class	**II 1 r**
Number of stars	**300**
Brightest	**10.7**
Constellation	**Cas**
RA	**23 57.0**
Dec	**+56 44**
Tirion	**3**
U2000	**35**
Description	**Cl, vL, vRi, vmC, st 11…18**

This is the type of deep-sky showpiece that used to reduce me to writing "WOW" and moving on. Take some time to study and enjoy this object. Then write out some notes that at least try to convey the sublime beauty of this cluster. It just takes practice. A study of NGC 7789 determined that there are 1000 members from 11–18 magnitude, so you won't run out of stars to observe anytime soon.

10×50 binoculars	S=6 T=6	Good site

10×—Bright, large, and a little brighter in the middle. Even at this low power, this compressed cluster will show two stars and the fuzzy background.

6 in.	f/6	S=6 T=6	Good site

25×—Bright, large, round cluster. 11 stars resolved across the sparkling face of this cluster. There is a hint of dark lanes within the cluster using averted vision.
100×—33 stars counted using 8.8 mm eyepiece, bright, large, little brighter in the middle, very compressed, rich. Stars sparkle on a huge, fuzzy background. Nice view in the RFT.

13 in.	f/5.6	5=7 T=9	Excellent site

60×—Bright, very large, rich and much compressed. There are 32 stars resolved and two dark lanes are obvious within the star cluster
100×—Resolved 76 stars. The cluster is about 1/3 of the field of view, with lots of beautiful chains of stars seen.
150×—The 14 mm Ultra-Wide-Angle (UWA) eyepiece provides the best view. I estimated over 160 stars resolved by counting 40 stars seen in the northwest quadrant of the cluster. This grouping now takes up about half the field and the lovely curving chains of stars are much more prominent. The dark lanes have excellent contrast at this power and are a striking feature of this beautiful cluster. These dark lanes wind through this group from edge to edge and give the impression of spiral structure. Higher powers, 220× and 300× will bring out some faint double stars within the cluster, but high magnification also looses some of the cluster aspect of this rich group of stars.

36 in.	f/5	S=6 T=8	Excellent site

165×—Very bright, extremely rich, very compressed. There are 75 stars resolved within the northwest quadrant, so the total number of stars seen across the face of this cluster is approximately 300! In the entire cluster I saw 4 orange stars and 20 that are dark yellow. Amazing dark lanes cut this cluster to pieces, then they wind their way out into the Milky Way. There are lots of stars just at the beginning of resolution with in the cluster, they create a foggy background that has an "oatmeal" texture to it.

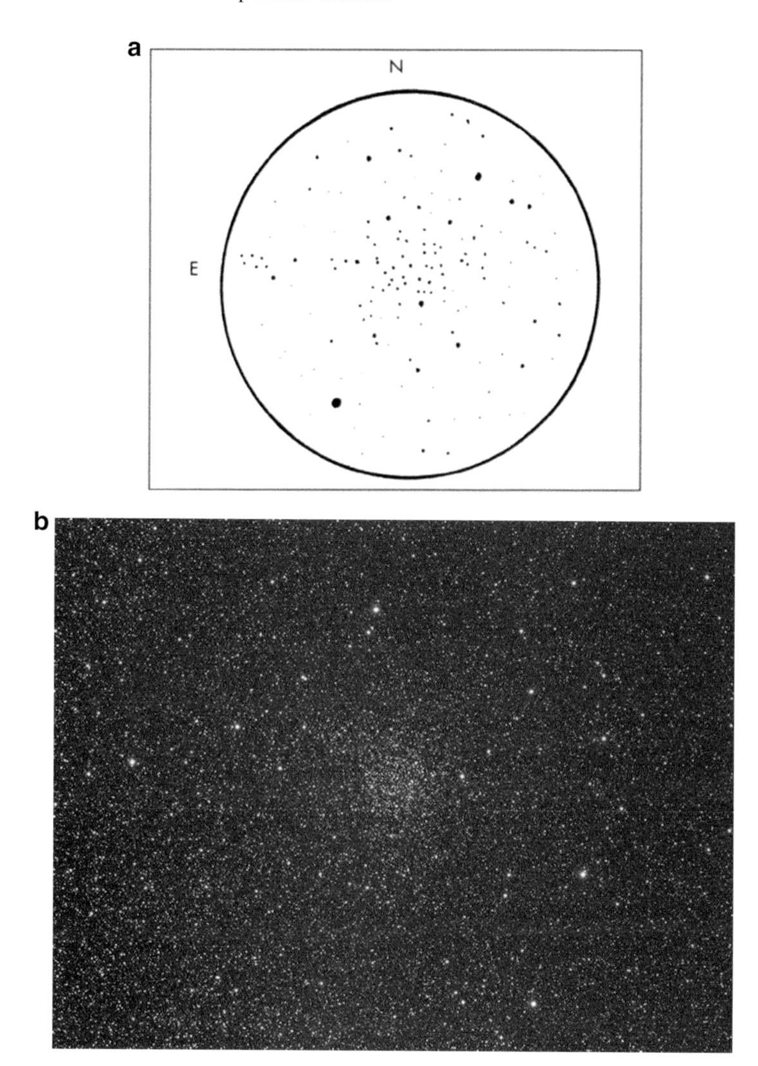

Fig. 13.16 (**a**) 6″ f/6; NGC 7789; FOV 50′; MAG 65×. (No, I am not going to draw NGC 7789 in a 36”). (**b**) AT90EDT SBIG ST8300M; *Photo: Dan Crowson*

Chapter 14

What Can Be Observed in Globular Clusters?

These tightly packed star clusters are indeed shaped like a globe, hence the name. They contain the most ancient stars within our Galaxy. As the Milky Way contracted, some concentrations of material formed in a halo surrounding the core. Stars formed within these huge concentrations of dust and gas, then those stars attracted one another and became the globular clusters. There are just over 100 globular clusters gliding around the center of the Galaxy, most of them located near the Core. Therefore, the constellations of Sagittarius, Scorpius, Centaurus and Ophiuchus are filled to the brim with these enchanting globes of stars.

There are several things other than shape which will distinguish a globular from an open cluster. At the eyepiece, you can see immediately that most globular clusters have a much more compressed core than any open cluster. This is because the globulars are much farther away from Earth than the open clusters, so the stars appear packed closer together. To grasp this approximation, wouldn't it be correct to approximate that every person on the island of Fiji is at the same distance from a given position in North America or Europe?

The same is true of the stars in a globular cluster; they really aren't packed right on top of one another, but it certainly looks like it.

The chemical composition of the stars in the two cluster types is also quite different. At the beginning of the Galaxy, it took time to create heavy elements in the cores of stars. Some of these stars exploded as a supernova and seeded the star material with elements more massive than helium. Because the stars in a globular are older, they do not include the heavier elements. Therefore, a spectral analysis is virtually always going to be able to quickly distinguish a globular from an open

S.R. Coe, *Deep Sky Observing*, The Patrick Moore Practical Astronomy Series,
DOI 10.1007/978-3-319-22530-2_14

cluster. What this means to you, at the eyepiece, is that the stars in a globular cluster are not going to show the wide variety of color that can be seen within galactic clusters. Thus the stars in a globular are generally white, though in a pretty large scope on a super night, some yellow or light-orange stars might appear in it.

Because globular clusters do indeed look like comets just starting to brighten up, Charles Messier was motivated to do a through job of finding the bright globulars available to an observer in Paris. Then William and John Herschel found many more as they swept the sky with their 18 in. telescope at the beginning of the nineteenth century. Therefore, any globular cluster that does not have an "M" or "NGC" designation is going to be a faint, difficult object.

Object	**NGC 2419**
Other names	**H I 218**
Type	**GLOCL**
Mag	**10.4**
Size	**4.1′**
Class	**2**
Constellation	**Lyn**
RA	**07 38.1**
Dec	**+38 53**
Tirion	**5**
U2000	**100**
Description	**pB, pL, lE 90°, vgbM**

This is a very distant globular cluster around the Milky Way. There was a time when it was thought that this small globe of stars might not be gravitationally bound to Our Galaxy. It has been more recently concluded that this distant object is still held within the grip of gravity by the Milky Way. Therefore, the nickname of "Intergalactic Tramp" no longer fits this globular cluster. Because of the distance of this object, its brightest stars are of 17th mag and are beyond all but the largest amateur telescope.

4 in.	f/8	S=7 T=7	Good site

90×—Faint, pretty small, round, brighter in the middle. Averted vision makes it larger. Obviously, there is no hint of resolution with the TV 102.
Trying 150× does not bring out any more detail on this pretty good night.

13 in.	f/5.6	S=6 T=8	Excellent site

150×—Pretty bright, pretty large, very little elongated 1.2×1 in PA 90, gradually much brighter in the middle. There are three levels of condensation and averted vision makes it grow larger, but no stars are resolved.
330×—Ragged edges, but still no stars resolved. There are four stars, all about 13th mag, on the four sides of the cluster, so it is boxed in by stars on all sides (See Fig. 14.1).

(continued)

(continued)

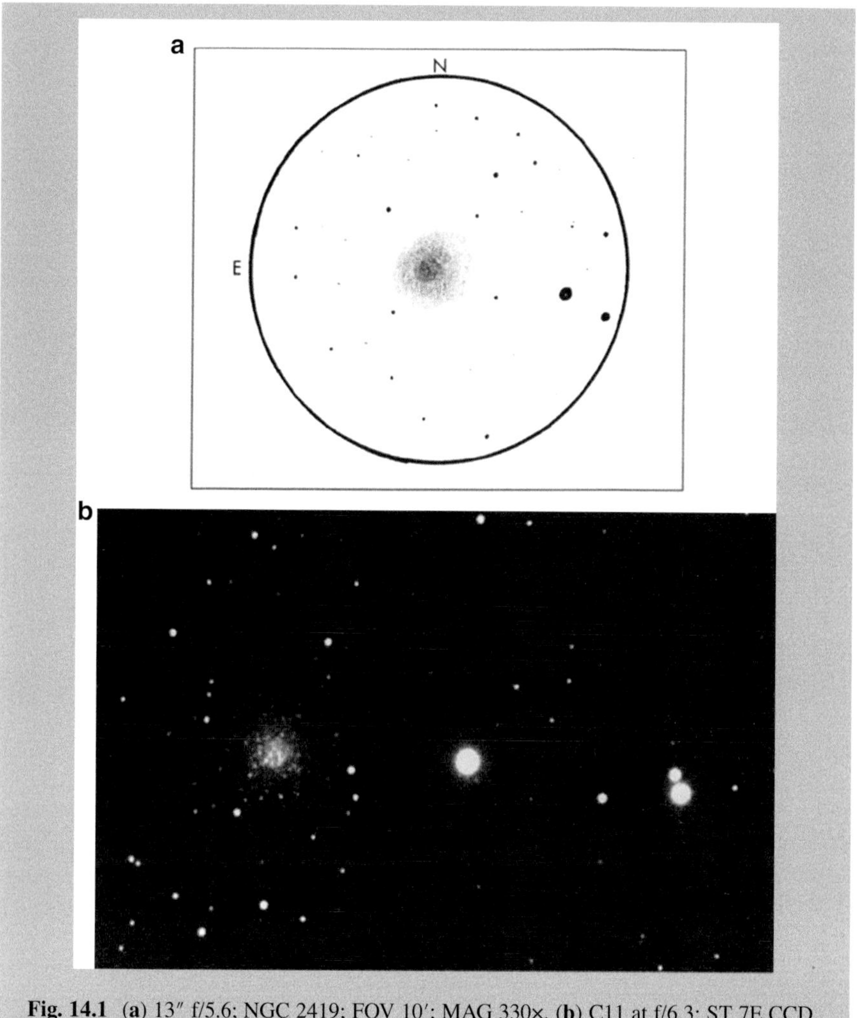

Fig. 14.1 (**a**) 13″ f/5.6; NGC 2419; FOV 10′; MAG 330×. (**b**) C11 at f/6.3; ST 7E CCD. *Photo: Robert Kuberek*

Object	**NGC 5139**
Other name	**Omega Centauri**
Type	**GLOCL**
Mag	**3.7**
Size	**36.3′**
Class	**8**
Constellation	**Cen**
RA	**13 26.8**
Dec	**–47 29**
Tirion	**21**
U2000	**403**
Description	**!!!eL, B, eRi, vvC**

Omega Centauri is an amazing globular cluster. This memorable object is at a far southerly declination, but does get above the horizon for Arizona observers. Notice two things from the data: first, the magnitude of this globular is 3.7, easily naked-eye; second, the NGC description starts with three exclamation points. This is a first-class object.

Naked eye S=7 T=8 Excellent site

1×—Easily naked-eye, pretty large even with no magnification, averted vision makes it grow larger. An unmistakable glow on a good night.

10×50 binoculars S=7 T=8 Excellent site

10×—Bright, very large, brighter middle. No stars are resolved with the binoculars, but the edges are ragged and the fact that this globular is oblate (not round) is immediately evident.

13 in. f/5.6 S=7 T=8 Excellent site

100×—Very bright, very, very large, extremely rich, very compressed, elongated 1.5×1 in PA 90. A myriad of stars, many form beautiful curved chains that loop around the core. Most of the stars are white, but a few are light yellow. Averted vision shows many faint, outlying stars beyond the central core section.

150×—This is a little too much power for an object that is only 10° above the southern horizon, but it is still an impressive view. A gigantic ball of stars.

12.5 in. f/6 S=7 T=6 Good site

140×—This globular fills the field with stars of a wide variety of magnifications. Many of the stars are white or off-white, but about 5 % are yellow or light orange, enough to lend a little color to the overall view. Magnificent chains of stars meander across the blazing core and then outward to the edge of the cluster. I chose an area that is about 10 % of the cluster space and count 52 stars in that area. This means that at this magnification, Omega Centauri is composed of approximately 520 stars and a bright, unresolved glow. It really seems like more stars than that, but figures don't lie. This observation is from Jim Barclay's backyard near Brisbane, Australia. As Omega transited almost straight overhead, we returned the scope to it several times and stared into the eyepiece with slack jaws at this fascinating collection of stars.

36 in. f/5 S=6 T=8 Excellent site

110×—Even with the 40 mm SuperWide eyepiece, Omega Centauri threatens to overflow this field of view with stars. It just fits into the eyepiece in the big scope. Obviously, there are lots and lots of stars. The core is a jumble of stars of several magnitudes, and this interlocking web of stars is well resolved. Behind these brighter stars is a finely grained sphere of fainter stars (Brian Skiff says like "grits" – we must be at the Texas Star Party). The spider web of stars is light yellow and the background is silvery white.

Fig. 14.2 20″ f/5; 20 min exposure. *Photo: Pierre Schwaar*

Object	**NGC 5694**
Other name	**H II 196**
Type	**GLOCL**
Mag	**10.2**
Size	**3.6′**
Class	**7**
Constellation	**Hya**
RA	**14 39.6**
Dec	**–26 32**
Tirion	**21**
U2000	**332**
Description	**cB, cS, R, psbM, r,* 9.5 sp 17.**

6 in. f/8 S=7 T=8 Excellent site

90×—Pretty bright, small, round and much brighter in the middle. Averted vision makes it larger and more prominent. It is somewhat mottled in moments of good seeing.

Using 170× does not bring out any more detail even on a very good night.

17.5 in. f/4.5 S=8 T=8 Excellent site
165×—Pretty bright, pretty large, round and brighter in the middle.
320×—Adding the Barlow lens will show six stars and a very grainy background. Averted vision makes this outer corona larger, but does not resolve any more stars.
This is a very distant globular and so it takes a steady, clear night and high magnification to resolve even a few of the member stars (See Fig. 14.3).

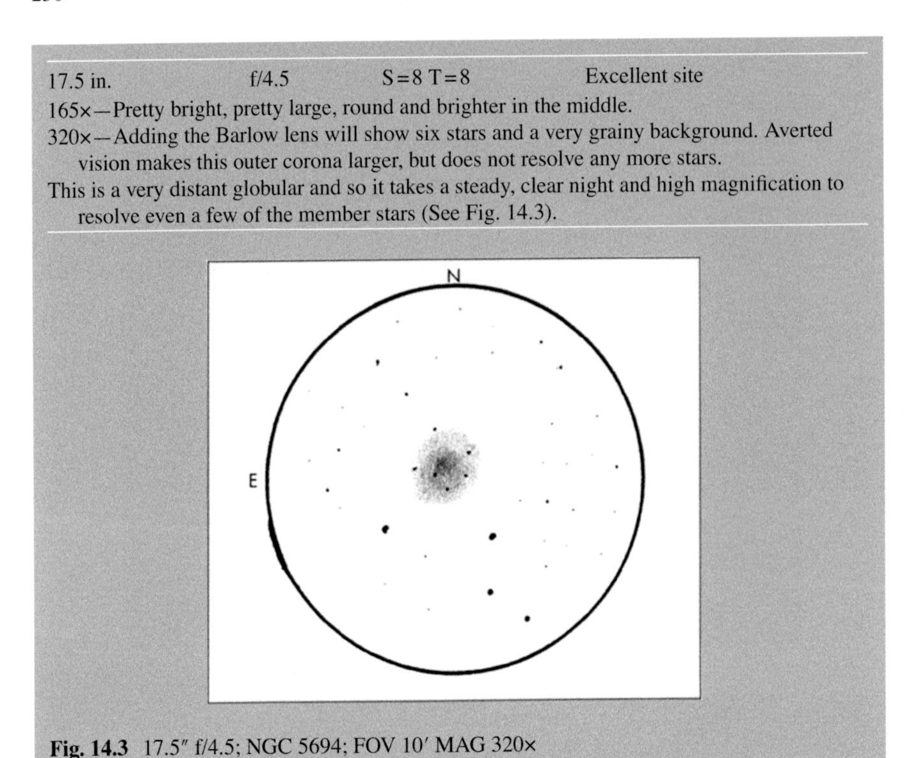

Fig. 14.3 17.5″ f/4.5; NGC 5694; FOV 10′ MAG 320×

Object	**NGC 5897**
Other names	**H VI 19**
Type	**GLOCL**
Mag	**8.6**
Size	**12.6′**
Class	**11**
Constellation	**Lib**
RA	**15 17.4**
Dec	**–21 01**
Tirion	**21**
U2000	**334**
Description	**pF, L, viR, vgbM, rrr**

Unlike so many globular clusters, this one does not exhibit a bright core. It has a pretty low surface brightness. If you are so used to seeing a bright Messier globular, this one will convince you that they are not all alike.

6 in. f/8 S=6 T=7 Good site
90×—pretty faint, large, irregularly round, gradually brighter middle, 6 stars resolved with direct vision and about 20 with averted vision. It has a pretty low surface brightness for a globular on a very good night.

17.5 in.	f/4.5	S=6 T=6	Good site

135×—Pretty faint, large, gradually brighter in the middle and not rich. I counted 17 stars resolved. This object needs to wait for a better night.

13 in.	f/5.6	S=8 T=10	Excellent site

150×—Pretty bright, large, irregularly round, pretty compressed, gradually brighter middle, 31 stars resolved, several in faint arms that extend out from main body.

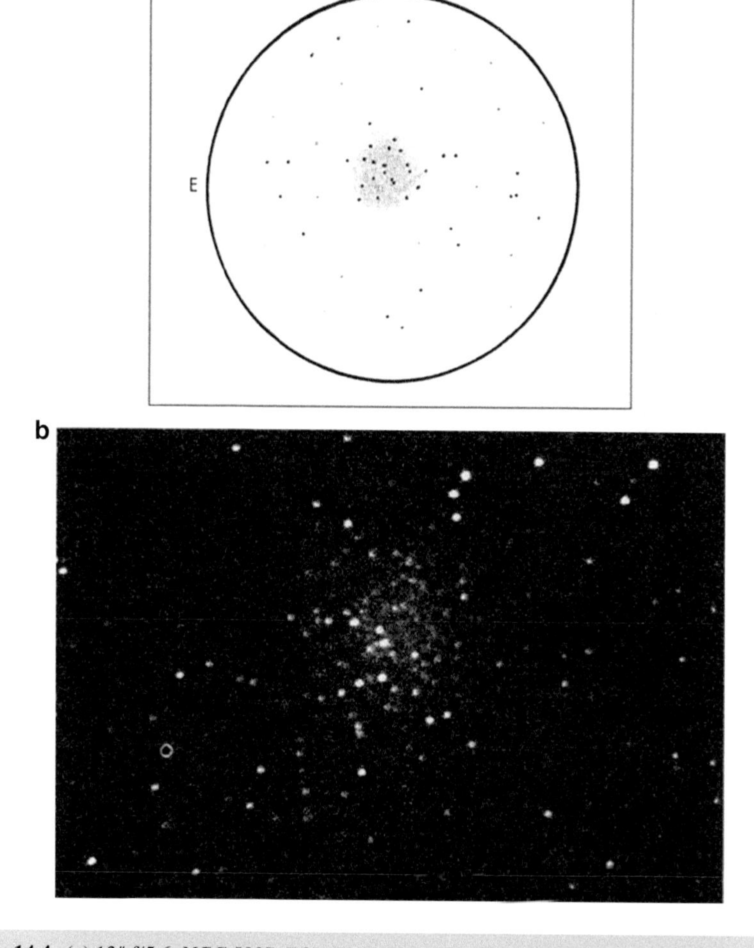

Fig. 14.4 (**a**) 13″ f/5.6; NGC 5897; FOV 20′; MAG 150×. (**b**) 12″ L×200 at f/7; ST 7E CCD; 6 min exposure. *Photo: Larry E. Robinson*

Object	**NGC 6144**
Other names	**H VI 10**
Type	**GLOCL**
Mag	**9.1**
Size	**9.3′**
Class	**11**
Constellation	**Sco**
RA	**16 27.3**
Dec	**–26 02**
Tirion	**22**
U2000	**336**
Description	**Cl, cL, mC gbM, rrr**

If you are making your way around the sky, looking at the Messier list, then you have probably had this object in your finderscope. It is located near to Antares and the bright globular M4 in Scorpius.

6 in. f/6 S=4 T=6 Mediocre site

100×—Pretty faint, large, very much elongated 4×1 and very little brighter in the middle, somewhat mottled in moments of good seeing.

Trying 150× does not bring out any more detail on this rather mushy night.

13 in. f/5.6 S=6 T=8 Excellent site

100×—Pretty bright, pretty large, compressed, three stars resolved, somewhat brighter in the middle.

150×—10 stars resolved, very grainy, many stars at the limit of the 13 in. on a good night, high in the mountains.

220×—Too much magnification; still 10 stars resolved, but very low surface brightness at high power. This globular cluster is at the edge of a very dark nebula.

36 in. f/5 S=6 T=7 Excellent site

200×—Pretty bright, compressed, somewhat brighter in the middle, round. I counted 39 stars resolved. Most stars are of the same magnitude, with a few brighter ones.

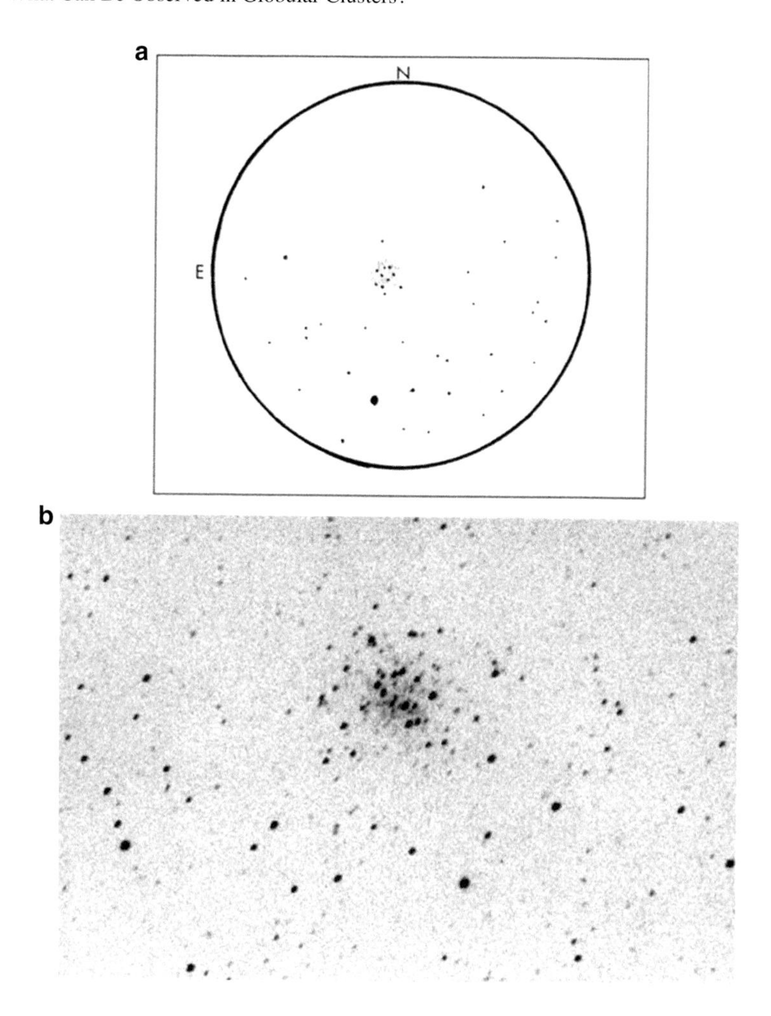

Fig. 14.5 (**a**) 13″ f/5.6; NGC 6144; FOV 20′; MAG 150×. (**b**) 12″ L × 200 at f/7; ST 7E CCD; 6 min exposure, *Photo: Larry E. Robinson*

Object	**M13**
Other names	**NGC 6205**
Type	**GLOCL**
Mag	**5.9**
Size	**16.6′**
Class	**5**
Constellation	**Her**
RA	**16 41.7**
Dec	**+36 28**
Tirion	**8**
U2000	**114**
Description	**!!eB, vRi, vgeCM,*11…**

I believe this globular gets a lot of press for several reasons: it is easy to find, it is one of the finest globulars, it is easily resolved in small scopes and it comes overhead for the Northern Hemisphere. Sir William Herschel estimated 14,000 stars in the cluster, and some hardy soul at Mount Wilson *counted* 30,000 on a plate from the 100 in. Hooker telescope in 1931. Actually, there are about half a million stars in M13. Poor 01' Charles Messier: his notes say "nebula, which contains no star". As Helen Lines used to say, "I envy Messier his skies, but not his telescopes."

When I was deciding on which bright objects to include in this book, I quickly decided on this one. The reason is that this is exactly the type of object that observers put in their telescopes over and over, then don't take good notes about how it appears in their telescope. This is not about my ego (OK, maybe a little bit), but notice how much you can write if you will get over it and spend the time to write down what you see.

Naked eye S=6 T=7 Excellent site

1×—M13 is just seen naked-eye, faint but there. As a comparison object, it has about half the size of the lagoon Nebula naked-eye. Averted vision does help.

11×80 finderscope S=6 T=7 Excellent site

11×—Bright, round, somewhat brighter middle. There are three layers of brightness, but the cluster is not resolved into stars.

6 in. f/6 S=5 T=6 Good site

100×—Bright, pretty large, brighter in the middle, compressed. There are 45 stars resolved and a giant fuzzy background. Averted vision doubles the size of this object. Seeing such a compact cluster with lots of space around it is fascinating; there is the effect of "Hey, where did all those stars come from?"

13 in. f/5.6 S=8 T=10 Superior site

60×—Very bright, very rich, gradually an extremely compressed middle, little elongated 1.2×1 in a PA of 90. This nice wide-angle view shows the two bright stars that guard M13 on either side and NGC 6207, a prominent galaxy to the north.

150×—Resolved 63 stars in northwest quadrant, so total resolved stars must exceed 250. Most of the stars are silvery white, but a few are light yellow and two are light orange. The two most prominent curved chains of stars loop away from the main cluster toward the northwest and southwest.

The dark "Propeller" feature is seen on the south side as small, thin dark lines. These three dark lanes divide the globular into unequal thirds, and there is a faint double star in the westernmost dark lane. The beautiful silvery sheen of this bright globular is most prominent at this magnification.

440×—Many, many faint pairs and groupings. Averted vision really "fills in" the cluster with a myriad of faint member stars. The dark lanes are more prominent at higher power. There is a chain of nine pretty bright stars that loop right across the core from east to west.

36 in. f/5 S=6 T=8 Excellent site

130×—As glorious as you might expect, hundreds of stars resolved with a fuzzy background of stars at the edge of resolution.

200×—Many, many beautiful chains, tendrils out from the main ball of stars, many double and triples stars within the curved chains that swirl out from the core. I counted 193 stars in the northwest quadrant of this globular, which means that at least 700 stars are resolved.

The dark "Propeller" feature is held with direct vision, but is more prominent with averted vision.

320×—Going to a 14 mm Meade UWA eyepiece shows off the Propeller better; there is more contrast. It also shows that there is another dark lane, across the core from the Propeller, that is also seen with direct vision. However, this higher power does not show any more stars. I know this is a showpiece that gets observed every time scopes are set up in the springtime, but the big scope provides a brand-new view of it. I had the feeling this was all that there was to see in M13 (See Fig. 14.6).

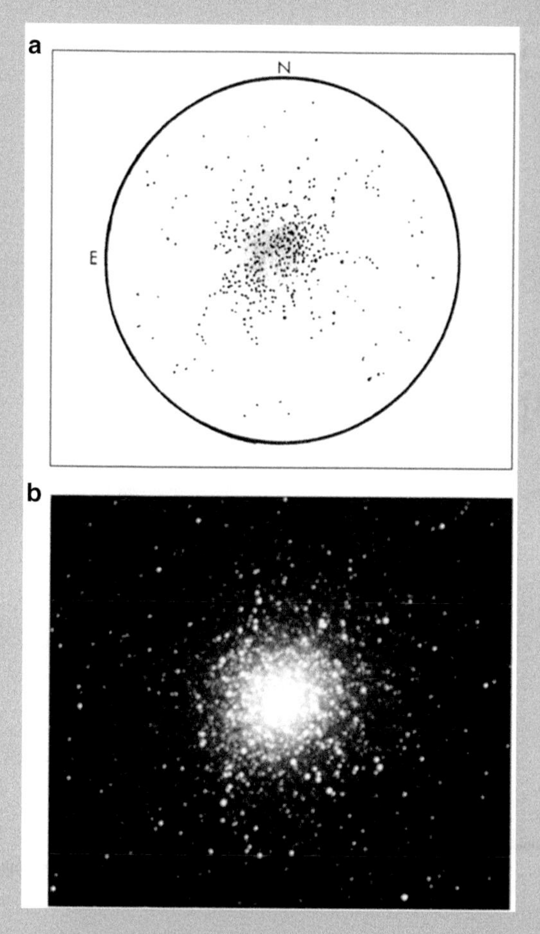

Fig. 14.6 (**a**) 13″ f/5.6; M13; FOV 10′; MAG 330×. (**b**) C11 at f/6.3; ST 7E CCD. *Photo: Robert Kuberek*

Object	**NGC 6229**
Other names	**H IV 50**
Type	**GLOCL**
Mag	**9.4**
Size	**4.5′**
Class	**4**
Constellation	**Her**
RA	**16 47.0**
Dec	**+47 32**
Tirion	**8**
U2000	**80**
Description	**vB, L, R disc, r**

William Herschel placed this globular in his planetary nebula category by mistake. Needless to say, he did not have spectral analysis to help him determine the category of each object on his list.

6 in. f/6 S=7 T=8 Excellent site

40×—Pretty bright, pretty small, much brighter middle, round.

100×—Bright core with one star to the north of the core, it was first seen with averted vision, then held steady.

135×—Still only the one star resolved in this globular, a bright core with edges that are not sharply defined. I see how this could be mistaken for a planetary nebula.

13 in. f/5.6 S=7 T=9 Excellent site

100×—Bright, pretty large, round, much, much brighter in the middle; no stars resolved, but cluster has three layers to brightness; averted vision makes it larger.

220×—A hint of resolution 10 % of the time.

330×—Five stars resolved, the brightest is north of the core of this globular, the other four fainter and to the south side.

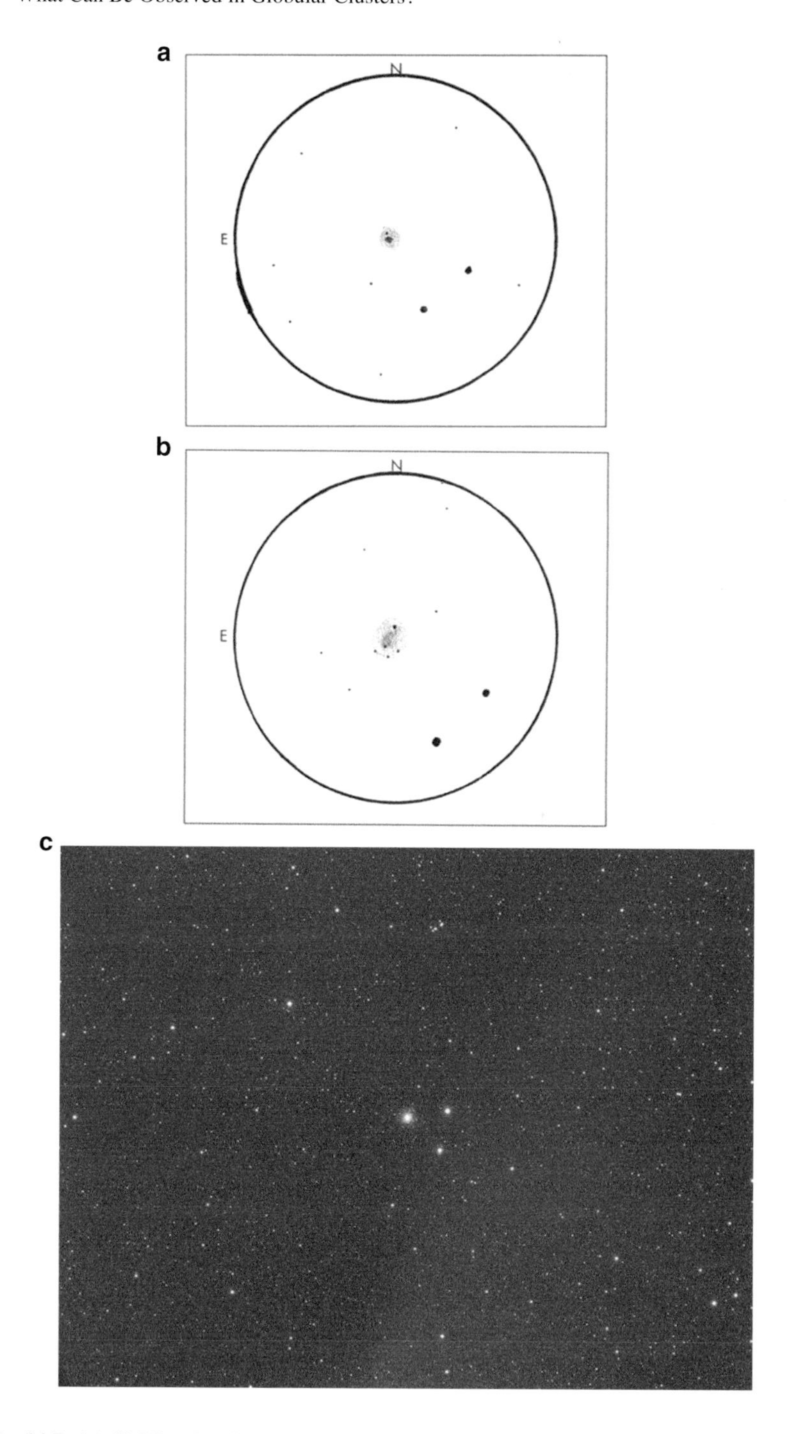

Fig. 14.7 (**a**) 6″ f/6; NGC 6229; FOV 30′; MAG 120×. (**b**) 13″ f/5.6; NGC 6229; FOV 10′; MAG 330×. (**c**) AT90EDT SBIG ST8300M; *Photo: Dan Crowson*

Object	**NGC 6366**
Type	**GLOCL**
Mag	**10.0**
Size	**8.3′**
Class	**11**
Constellation	**Oph**
RA	**17 27.7**
Dec	**–05 05**
Tirion	**15**
U2000	**248**
Description	**F, L, vlbM**

6 in. f/6 S=7 T=7 Good site

40×—Just barely seen, never held steady with direct vision.

65×—Helps some, but still extremely faint; pretty large, round, only seen about 50 % of time; averted vision helps.

13 in. f/5.6 S=6 T=8 Excellent site

150×—Faint, large, round, not brighter in the middle.

220×—Counted 11 stars when I raised the power on this low-surface-brightness globular. It is somewhat strange for a globular because it is not compressed at all. It looks like a pretty faint open cluster with several brighter members. When a friend looked into the eyepiece, he said it looked like oatmeal. I agreed.

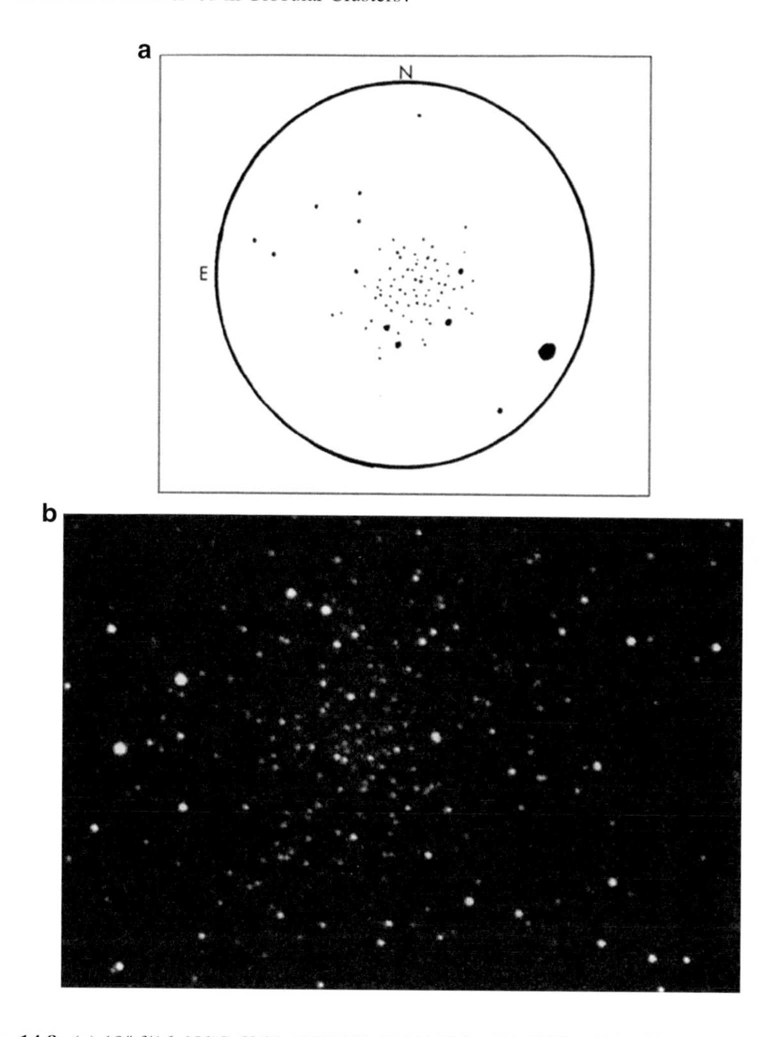

Fig. 14.8 (**a**) 18″ f/16; NGC 6366; FOV 20′; MAG 135×. (**b**) 12″ L×200 of f/7; ST 7E CCD; 6 min exposure. *Photo: Larry E. Robinson*

Object	**NGC 6517**
Other names	**H II 199**
Type	**GLOCL**
Mag	**10.3**
Size	**4.3′**
Class	**4**
Constellation	**Oph**
RA	**18 01.8**
Dec	**–08 58**
Tirion	**15**
U2000	**294**
Description	**pB, pL, R, rr**

The field of view near this globular cluster is very star-poor and this dark field may exhibit so much extinction that I am not able to resolve any stars on the face of this globular.

6 in.	f/8	S=6 T=7	Good site

90×—Faint, pretty small, round, no stars are resolved and very little brighter middle. The cluster has few stars in its field of view, so it must be obscured by dust. It is low surface brightness and it not much in the 6 in. even on a good night. Higher powers do not resolve the cluster into stars.

13 in.	f/5.6	S=7 T=8	Excellent site

100×—Pretty bright, pretty small, round, much, much brighter middle; easy to see as non-stellar.

220×—About all the power this globular can take; a little fuzzy at the edges; averted vision makes it grow. Still no resolution into stars. The core is elongated 1.5×1 in PA 45.

330×—No resolution, but very grainy at the edges.

It seems that just a little more aperture or a really terrific night at high altitude might show a few of the stars as tiny points of light (See Fig. 14.9).

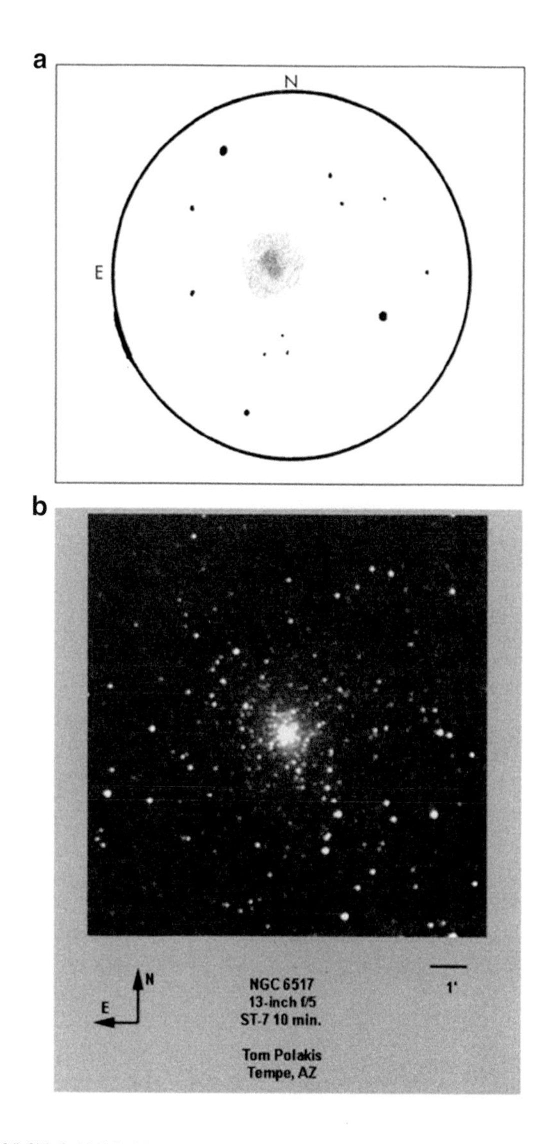

Fig. 14.9 (**a**) 13″ f/5.6; NGC 6517; FOV 10′; MAG 330×. (**b**) 13″ f/5. *Photo: Tom Polakis*

Object	**NGC 6652**
Type	**GLOCL**
Mag	**8.9**
Size	**3.5′**
Class	**6**
Constellation	**Sgr**
RA	**18 35.8**
Dec	**–32 59**
Tirion	**22**
U2000	**378**
Description	**B, S, lE, rrr, st15**

6 in. f/8 S=6 T=7 Good site

90×—Pretty bright, pretty small, little elongated 1.2×1, much brighter middle, there are three levels of brightness toward the center. There is a star of about 11th magnitude at the western edge. 170× now there are three other stars resolved and a stellar core is seen, averted vision makes it larger.

13 in. f/5.6 S=6 T=7 Good site

150×—Pretty bright, pretty small, pretty much compressed, much brighter middle, elongated 1.5×1 in PA 0.

220×—Three stars are resolved and there are grainy edges to this globular, all in a rich field of view.

330×—Three stars are held steady, another two or three come and go with the seeing—they literally wink on and off. Averted vision makes a very large difference, doubling the size of this object.

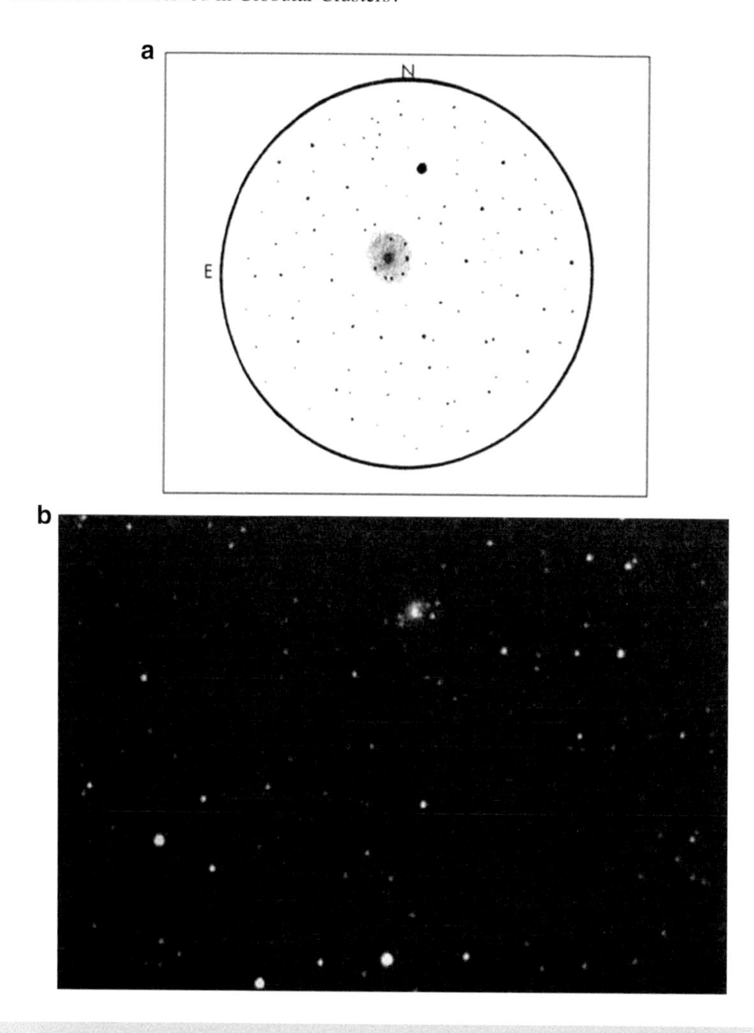

Fig. 14.10 (**a**) 13″ f/5.6; NGC 6652; FOV 20′; MAG 220×. (**b**) 12″ L × 200 at f/7; ST 7E CCD; 6 min exposure. *Photo: Larry E. Robinson*

Object	**NGC 6712**
Other names	**H I 47**
Type	**GLOCL**
Mag	**8.2**
Size	**7.2′**
Class	**9**
Constellation	**SCT**
RA	**18 53.1**
Dec	**−08 42**
Tirion	**15**
U2000	**295**
Description	**pB, vL, irr, rrr**

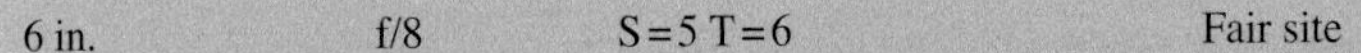

6 in.	f/8	S=5 T=6	Fair site

140×—Pretty bright, pretty large, irregularly round, very little brighter in the middle, 2 stars resolved with direct vision and averted vision adds another 12 or so. This cluster sparkles with stars at the limit of the 6 in. scope.

13 in.	f/5.6	S=7 T=8	Excellent site

100×—Pretty bright, large, much compressed, brighter middle. eight stars resolved, averted vision really makes it grow to about double in size.

150×—Fourteen stars resolved, the last four extremely faint. Great view of an extremely compressed cluster through a very rich Milky Way foreground.

220×—Twenty stars counted; last five difficult, with averted vision only. This cluster is made up of the finest stardust with many faint and very faint members, which just show themselves in good seeing. Somewhat triangular in shape with direct vision and round with averted vision, a bizarre change in shape depending upon where your eye is looking (See Fig. 14.11).

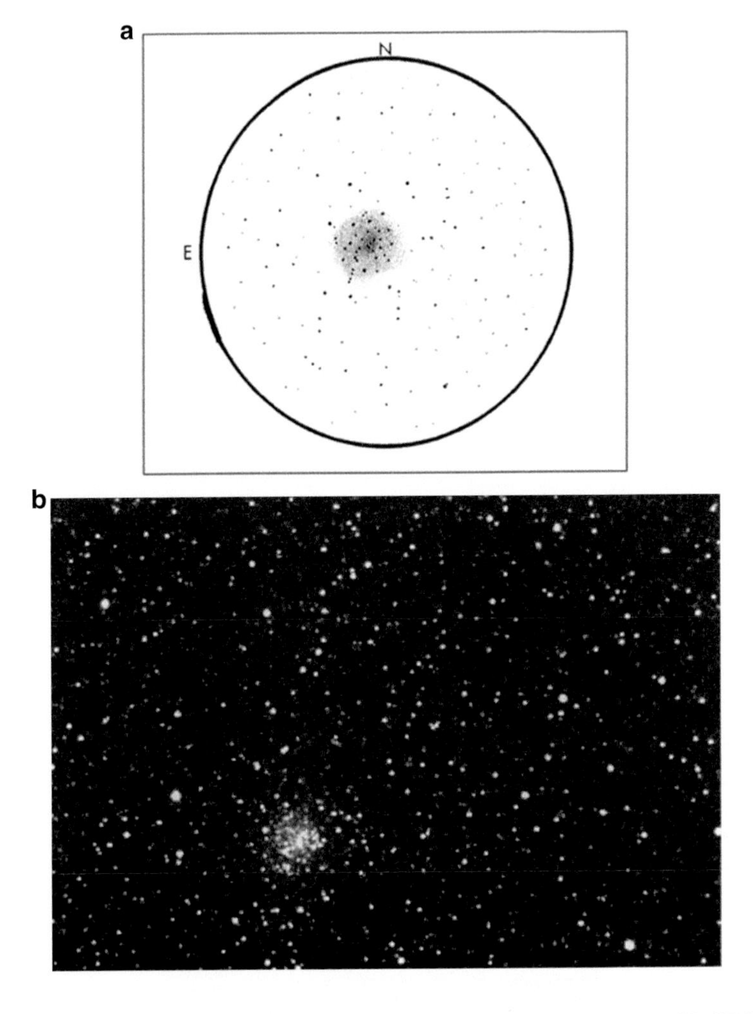

Fig. 14.11 (**a**) 13″ f/5.6; NGC 6712; FOV 25′; MAG 150×. (**b**) 12″ L×200 at f/7; ST 7E CCD; 6 min exposure. *Photo: Lorry E. Robinson*

Object	NGC 6723
Other names	Dunlop 573
Type	GLOCL
Mag	7.3
Size	11.0′
Class	7
Constellation	Sgr
RA	18 59.6
Dec	–36 38
Tirion	22
U2000	378
Description	vL, vIE, vgbM, rrr

10×50 binoculars	S=6 T=7	Good site

10×—Easy to spot as bright area, little brighter middle.

13 in.	f/5.6	S=6 T=7	Good site

135×—Bright, large, round, much brighter in the middle, 30 stars resolved at the edges of this big globular. There are several long chains of stars that wind their way out from the edges.

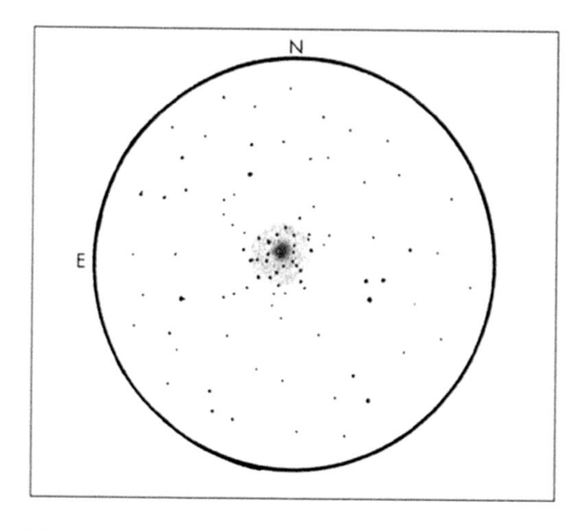

Fig. 14.12 13″ f/5.6; NGC 6723; FOV 20′; MAG 150×

Object	NGC 6934
Other names	H I 103
Type	GLOCL
Mag	8.9
Size	2′
Class	8
Constellation	Del
RA	20 34.2
Dec	+07 24
Tirion	16
U2000	209
Description	B, L, R, rrr,*16...

This is the type of object that responds with a much better view on a clear, transparent night. On a night I rated 6/10 for seeing and transparency, down on the floor on the desert near the Organ Pipe Cactus National Forest, I could only resolve three stars with a mottled core at 200×. At the same power on a beautiful 9 out of 10 night in the Red Rock country near Sedona at high altitude, this globular blazed with 40 stars resolved, 6 of them in the core area. This is the kind of observation to postpone until those rare evenings when stars twinkle very little and the Milky Way lights up the sky.

6 in. f/6 S=7 T=7 Excellent site

25×—Just seen as non-stellar.

65×—Pretty bright, pretty small, round, somewhat brighter middle; there are two layers of increasing brightness toward the middle.

100×—No resolution, but very grainy, averted vision makes it larger.

13 in. f/5.6 S=7 T=8 Excellent site

100×—Bright, pretty large, round, much compressed, round, much, much brighter middle. There are three layers of brightness toward the core, but no star is seen at low power. Averted vision makes it much larger, at least double the size with direct vision.

330×—Resolved 21 stars at high power, including 6 in the core.

There is a nice 8th mag star to the west of this globular. It serves two purposes: one, a good star to use for getting exact focus; two, its size can provide information concerning the seeing. A large Airy disk means the seeing is not good enough to try to resolve distant globular clusters.

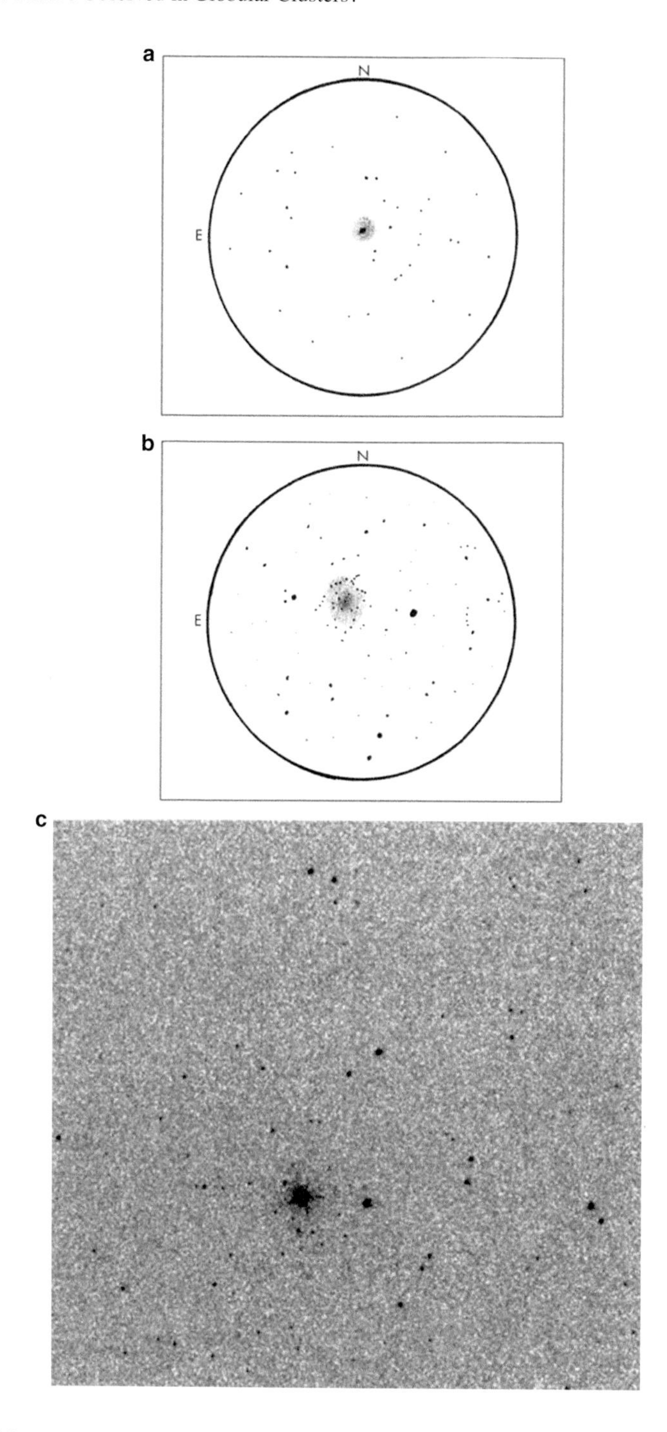

Fig. 14.13 (**a**) 6″ f/6′ NGC 6934: FOV 40′; MAG 100×. (**b**) 13″ f/5.6; NGC 6934; FOV 20′; MAG 330×. (**c**) 12″ L×200 at f/7; ST 7E CCD; 6 min exposure *Photo: Larry E. Robinson*

Object	**NGC 7006**
Other names	**H I 52**
Type	**GLOCL**
Mag	**10.6**
Size	**2.8′**
Class	**1**
Constellation	**Del**
RA	**21 01.5**
Dec	**+16 11**
Tirion	**16**
U2000	**209**
Description	**B, pL, R, gbM**

This is the other famous globular that is very far away from the Earth. While viewing this deep space wanderer, think of the view of Our Galaxy you would have from 180,000 light years away.

13 in. f/5.6 S=7 T=8 Excellent site
100×—Pretty bright, pretty large, round, pretty gradually much brighter middle.
330×—There are three layers of brightening toward the core; the globular is very grainy. However, there is not resolution into stars, with either averted or direct vision.

18 in. f/6 S=7 T=8 Excellent site
210×—Three stars are superimposed on the surface of NGC 7006. One was held steady, while the other two appeared and disappeared with the seeing. You are observing some faint stars when they are at the limit of 18 in. of aperture on a good night.

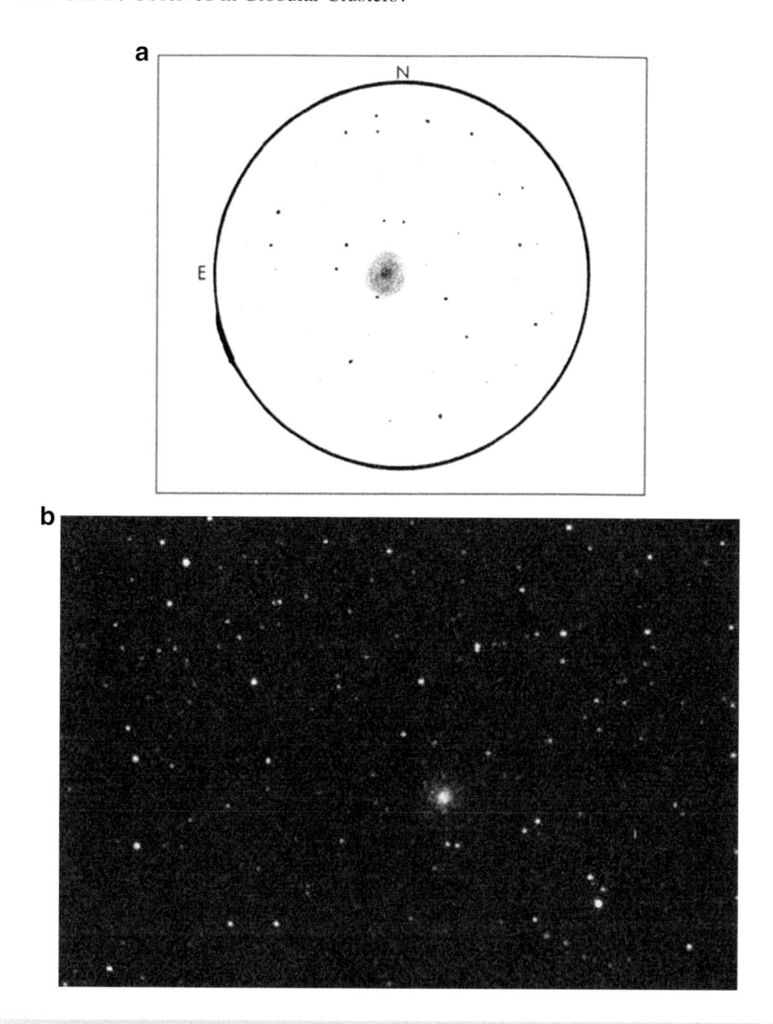

Fig. 14.14 (**a**) 13″ f/5.6; NGC 7006; FOV 10′; MAG 330×. (**b**) 12″ L × 200 at f/7; 6 min exposure. *Photo: Larry E. Robinson*

Chapter 15

Why Would I Want to Use Binoculars to View the Sky?

Many advertisers shout about the high magnifications available using their telescopes. There is no doubt that high magnifications yield beautiful views of the Moon and planets and fine detail in some deep-sky objects. However, many objects in the sky are too large to fit into the field of view of a high-power eyepiece. These objects demand a wide field of view to appreciate their beauty and delicate form (See Fig. 15.1).

There are several ways to view large areas of the sky. The simplest is naked eye. Your eye, when properly dark adapted, is a marvelous instrument. However, stepping up to a pair of binoculars will help a lot. The added light-gathering power of the lens system in a pair of binoculars will allow you to see much fainter celestial objects. Along these same lines, a finderscope on most telescopes is a monocular (half a binocular). This small refractor can provide some excellent rich-field views.

The biggest advantage of binoculars is that they are easily portable. What can be simpler than carrying a pair of binoculars out to an observing site? Any set of optics that can be carried on a bicycle with ease has the portability edge. The disadvantage of binoculars is aperture. Even at their largest, binoculars are not going to show the faintest details which can be seen in an amateur telescope. It is impossible to build an auto-mobile that will get good gas mileage and win a Grand Prix race. Sorry, you just can't have it all.

One other advantage to binoculars is that they are worth having for a variety of uses, other than astronomy. Taking a pair of binoculars to a sporting event can put you right in the action. At a concert, you can see all the performers, close up. If you are traveling to a park or other tourist spot, then some optical aid will allow you to see lots of great detail. Looking at the multicolored walls of the Grand Canyon or relief carvings in a cathedral, a good pair of binoculars can allow you to see much more than naked eye (See Fig. 15.2).

S.R. Coe, *Deep Sky Observing*, The Patrick Moore Practical Astronomy Series,
DOI 10.1007/978-3-319-22530-2_15

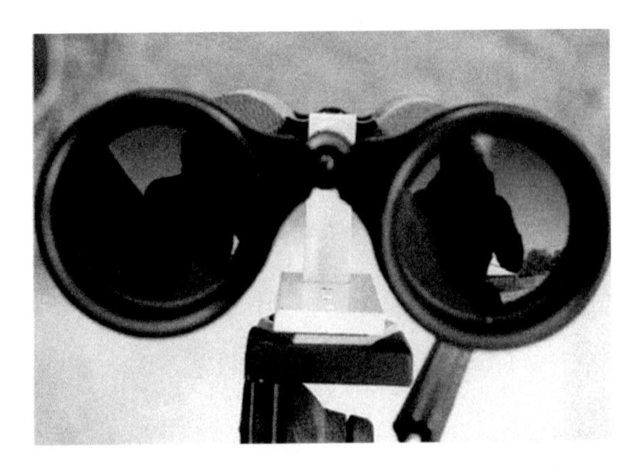

Fig. 15.1 Binoculars are a very easy to use optical aid for observing the sky

Fig. 15.2 Binoculars come in a variety of sizes; *left* to *right*: 11×80, 10×50, 8×25

Binoculars are designated by two numbers: magnification and then aperture in millimeters. Common size ranges for binoculars are: 7×35, 7×50, 10×50, 11×80, 15×80 and 20×80. So, a 7×50 (spoken "seven by fifty") binocular has eyepieces that yield seven times magnification and have a pair of 50 mm lenses at the front objective. In these sizes they are quite affordable and easy to transport. Moving to binoculars with an aperture of 100 mm and up will raise the price considerably. For those reasons, let's stick to the convenient sizes for this discussion.

If you are looking for the widest field possible, then a pair of 7×35 or 7×50 binocs will give a field generally between 5° and 10° in size. This is about the size of the Bowl of the Big Dipper or larger than the Belt and Sword of Orion. If you are trying to view most of a large constellation or all of a small one, these low-power

Fig. 15.3 Binoculars give a wide field of view of many spectacular regions of the sky

binocs are just what you want. With a pair of smaller binoculars, such as 8×25, you can get a very wide field, but these small binoculars don't have the light grasp that a larger pair will have (See Fig. 15.3).

Getting a higher-power pair of binoculars will show you dimmer stars and fainter deep-sky objects. So the 10×50 and larger binoculars are great for looking at nebulae and resolving star clusters. The larger and more powerful binoculars will also show more detail in galaxies (See Fig. 15.4).

In magnifications of 10× and above the weight of the binocs becomes a factor to consider. As your arms tire, it becomes difficult to hold them steady enough for the kind of view they are capable of delivering. For this reason it is wise to consider some kind of support for high-power binoculars. Many varieties of binocular supports have been described and sold in astronomy magazines over the years. If you don't feel comfortable constructing your own stand then there are several types of tripods available through advertisers in magazines or at better camera stores in your area. Most large, modern binocs come with a tripod adapter which has a standard thread that fits the tripod head. This will allow you to easily mount the binocs on a sturdy support. The support will provide a field of view which will not jiggle when you observe with tired, cold hands.

Fig. 15.4 A tripod will steady heavy binoculars for a better view

This is a deep-sky book, but comets also provide an excellent target for binoculars. Because comets will occasionally form a tail several degrees in length, they often lend themselves best to wide-field instruments. There were several nights of interesting viewing when Comet Bradfield passed near M10 and M12 in Ophiuchus.

Binoculars were essential for watching Comet IRASAraki-Alcock as it moved across the sky in Draco. This comet passed so close to the earth, that even in the 10 × 50 s it would move against the star background as you observed!

Both of the bright comets that were seen in the 1990's, Comet Hyakutake and Comet Hale–Bopp, were magnificent in binoculars. They showed off bright tails and fascinating detail around the nucleus that changed from night to night. Comet Hale–Bopp passed near the bright star cluster M 34 in Perseus one night. The view was unique and enthralling as this big, bright comet glided in front of that compressed ball of stars (See Fig. 15.5).

Dave Fredericksen, Chris Schur and I had the good fortune to travel to Australia in 1986 to view Comet Halley. We spent a week in Jim Barclay's backyard near Brisbane and marveled at the southern sky. Taking binoculars was easy and they were on hand for convenient viewing of unfamiliar parts of the Milky Way. The 10 × 50 s were great on the Eta Carina Nebula, the Coalsack and the Magellanic Clouds. However, I will never forget Comet Halley in a pair of 15 × 80 binocs. We happened to be observing the comet while it was quite active. The gas tail was very prominent and there were several layers of brightness to the coma. The tail changed its orientation and detail came and went from one night to the next. The big binoculars provided a front-row seat to all the action (See Fig. 15.6).

Fig. 15.5 Comet Hale–Bopp and M-34 in Perseus

Fig. 15.6 The Coalsack—a dark nebula near the Southern Cross

I have had the good fortune to use several styles of binoculars over the years. The simplest way to give you some information about their performance is to provide you with some observations of wide-field objects. Then you can decide which type of optical system meets your needs. There will be a simple set of information about each object and then the observation with binoculars.

M31

Sb galaxy 00 41.8+41 16 4.0mag 160′×40′ The Andromeda Galaxy is bright enough to be seen naked-eye from even a mediocre site. It starts to show off its grandeur in medium binoculars from a good site. With my 10×50 s it is bright, large, and elongated 3×1. The core is seen easily and averted vision makes it much larger. The dark lanes are not visible in my 10×50 binocs. Moving up to a pair of 11×80 s makes a big difference. The big binoculars frame the Andromeda Galaxy nicely and the galaxy is two fields of view long (about 5°) on a night I rated 7/10. The dark lanes are evident on a good evening and the companion galaxies can be seen with ease. Within the southern arm a bright H II region, NGC 206, is visible in the big binoculars and the rest of the spiral arms sparkle with mottling.

NGC 869

OC 02 19.0+57 09 4.3mag 30′ **NGC 884** OC 02 22.4+57 07 5.1mag 29′ The Double Cluster in Perseus is one of the most famous objects that did not make Charles Messier's list. It can be seen naked-eye as a bright spot in the Winter Milky Way. The following binocular observations are for both clusters together.

Fig. 15.7 The Andromeda Galaxy (M 31). 300 mm lens; Canon T2i camera; 5 min exposure

Fig. 15.8 The Double Cluster. ED 80 at f/6; Canon T2i camera; 4 min exposure

10×50—Bright, very large, pretty compressed, 13 stars resolved, several yellow stars and one orange that is between the clusters.

11×80—Very bright, very large, compressed, 18 stars resolved, two orange stars and several yellow. With averted vision approximately 12 stars pop out from the fuzzy background. It is a fascinating effect to go from direct and then averted vision to see these stars and then have them go away with direct vision. Several lovely curved chains of stars wind their way out into the Milky Way from the edges of the clusters.

Mel 20

Per OC 03 22.0+49 00 1.2 185 The Alpha Perseus Association is a huge, nearby open cluster. Even with the little 8×25 binoculars there are 16 stars resolved, most in a lovely, long, looping chain of pretty bright stars. The 10×50 s resolved 39 stars, 6 pretty bright and the rest pretty faint. There are several triple stars and the beautiful loop of stars to the south of Alpha Per is composed of 12 stars. The cluster is over half of the field of view. Even though the 11×80 binoculars resolve 44 stars, they don't add much to what was seen with the 10×50 s.

M45

OC 03 47.0 24 07 2mag 120′ The Pleiades are among the best-known objects in the sky beyond the Solar System. Observing naked eye, six stars are seen even from my suburban backyard. The best I have ever done from a dark-sky site is a dozen stars, with the last two stars being averted vision only. Using the 8×25 binoculars,

Fig. 15.9 The Alpha Perseus Association. 200 mm lens; Canon T2i camera; 3 min exposure

Fig. 15.10 The Pleiades (M 45). 300 mm lens; Canon T2i camera; 8 min exposure

I can resolve 24 stars in the cluster; the chain of stars which trails off to the south is easily seen. Going to the 10×50 binocs shows off 36 stars in a cluster that takes up the central half of the field of view. There is a wedge of nebulosity with Merope at the head and a little round fog surrounding two other Pleiades stars. The 11×80 s are the perfect optical aid for this object. There are 64 stars resolved and the nebulosity is easy around all the bright stars, except Electra. The size of these nebulae grows with averted vision. The small triangle of stars within the Pleiades is resolved and there are many star pairs in the cluster. What an excellent view of this famous cluster!

Mel 25

OC 04 27.0+16 00 1.0mag 330′ The Hyades are an even larger cluster than the Pleiades and even though Aldebaran is not a member, it does add sparkle. On this huge grouping, the 8×25 binoculars are an excellent tool. The cluster fills about 60 % of the field and 47 stars are resolved, including many pairs in this nice view. Moving up to the 10×50 s will resolve 55 stars and show that there are not many faint members in the Hyades. Now the cluster aspect is almost lost because the field of view is filled with stars. Aldebaran is a nice orange color and there are lots of nice star pairs, but no chains of stars.

Fig. 15.11 The Hyades. 85 mm lens; Canon T2i camera; 1 min. exposure

M42

EN 05 35.4−05 27 6mag 66′×60′ The Orion Nebula region is glorious in a good pair of binoculars on a dark night. With the 10×50 binoculars the entire Sword of Orion fits in the field of view. The separation between M42 and M43 is easy and the fact that M43 is shaped like a thick comma is easy to see. The outer reaches of M42 are also easy and they appear as "wings" or a "flame" that is sprouting from the bright section around the central Trapezium stars. The 11×80 binoculars will show 11 stars involved within the nebula, other than the easy Trapezium stars. The central region has some texture or mottling. The fainter, outer portions of the nebula, near lota Ori, are just seen.

Cr 70

OC 05 36.0−01 00 0.4mag 150′ This is one of those deep-sky objects that you might not know even had a designation. Collinder 70 is the star association that includes the stars of the Belt of Orion. Using the 8×25 binoculars will show off 22 stars in beautiful loops and chains of stars, all around the three bright stars of the Belt. With the 10×50 s there are 53 stars counted and the view is vastly improved. Now the chains are filled with many faint stars and many of the pretty bright stars are blue–white in color. A few yellow–orange stars are also included among the group. The 11×80 binoculars show too little of the sky to allow a good view of the association.

Fig. 15.12 The Orion Nebula (M 42). ED 80 at f/7; Canon T2i camera; 5 min exposure

M35

OC 06 08.9 + 24 20 6mag 29′ (See Fig. 15.7).

This star cluster in Gemini is just seen naked-eye from a dark site. The 10 × 50 binoculars resolved 20 stars within the cluster and showed a fuzzy background of unresolved stars. There is a nice light-orange star near the center of the cluster. It stands out well from the winter Milky Way. With the 11 × 80 binoculars, I can resolve 39 stars in this beautiful cluster. NGC 2158 is a cluster near M35, just seen as a faint smudge. With the big binoculars, M35 really stands out from the background of the Milky Way. The orange star near the middle of the cluster is easy and there are two other light-yellow stars within the cluster.

NGC 2244

OC 06 32.4 + 04 52 4.8mag 24′ The Rosette Nebula in Monoceros is a faint glow in the Milky Way to the naked eye. The coarse cluster within this wreath of nebulosity is shown in 10 × 50 binoculars but I have never seen the nebula clearly at this power. However, the 10 × 50 s do show 20 stars, the brighter ones arranged in two parallel rows. With the larger 11 × 80 s there are 33 stars resolved and a hint of the nebulosity is seen as a wide, faintly glowing arc around the star cluster.

M44

OC 08 40.1 + 19 59 4mag 90′ The Beehive is aptly named. Even with the small 8 × 25 binoculars, there are 11 stars resolved and the cluster is bright and large. The 10 × 50 s will show 27 stars, with several star pairs seen. This large cluster is about

Fig. 15.13 M35 in Gemini; ED 80 at f/6; 3 min exposure

Fig. 15.14 The Beehive cluster (M 44). ED 80 at f/6; Canon T2i camera; 2 min exposure

one-third of the field. The 11 × 80 binoculars will resolve 44 stars in the Beehive swarm. There is one triple star, one member of which is yellow. The cluster includes a light-orange star within the somewhat scattered grouping.

M6

OC 17 40.1 − 32 13 4.2mag 33′ This beautiful star group has been called the Butterfly Cluster. There are two lovely curved chains of stars that form the antennae and the brightest stars from the wings of a delicate celestial butterfly. There is also an orange star (BM Sco) at one wingtip. With the 10 × 50 binoculars there are 14 stars resolved. This group is much more compressed than nearby M7. The orange star is pretty easily seen. Moving up to the 11 × 80 s the variable star BM Sco is easily seen as light orange. There are 20 stars resolved and the "butterfly" shape is easily seen. The cluster is about one-fourth of the field of view.

M7

OC 17 53.9 − 34 49 3.3mag 80′ (See Fig. 15.15).

This very large and somewhat scattered star cluster in Scorpius is near the stars which form the stinger. As a matter of fact, the wide field of the small 8 × 25 binoculars can accommodate both clusters M6 and M7 and the two stars of the stinger. There are eight stars resolved within M7. Moving up to either the 10 × 50 or 11 × 80 binoculars will resolve 15 stars with either aperture. Not many faint stars are within M7.

Fig. 15.15 M6 and M7 in Scorpius; 200 mm f/3.5 lens; 15 min exposure

M8

OC+CL 18 04.3−24 20 4.6mag 55′ The Lagoon Nebula got its name from the dark nebula that appears to form a lagoon with the glowing nebulosity. From even a moderately dark site, it is easily seen naked-eye and elongated E–W. With the small 8×25 binoculars, there are four stars resolved within the nebula. The faint glow of the nebula is small, but it is there. The 10×50 s show the nebula much larger and resolve eight stars in that glow. The nebula is much brighter in the center. Moving up to the 11×80 binoculars will bring out 14 stars with direct vision and add another 6 to 10 with averted vision. Now the nebula is obvious and the bright central region is seen as elongated with a thin dark lane within it.

M11

OC 18 51.1−06 16 5.8mag 25′ (See Fig. 15.17).

This very compressed open cluster shows the limitations of binoculars very well. It is easy to see the pretty bright glow of the cluster and distinguish it from the background. However, it is not easy to see individually resolved stars in the cluster. Even the little 8×25 binoculars can easily see this cluster as a detached spot on the Scutum Star Cloud. With the 10×50 binoculars a bright middle is seen and about 30 % of the time one star is resolved. With the 11×80 s there are three stars detected, one of them faint. All these binoculars show lovely dark lanes spreading out from the glow of M11.

Fig. 15.16 The Lagoon Nebula (M 8); ED 80 at f/6; Canon T2i camera; 4 min exposure

Fig. 15.17 The summer milky way is a prime binocular viewing area

Fig. 15.18 The coat hanger

Cr 399

OC 19 25.4+20 11 5.2mag 8.0′ (See Fig. 15.18).

This is Collinder 399, also known as the Coathanger. Next clear night, go look for yourself. With the 8×25 binocs there are 13 stars seen, the brightest 7 that form the coathanger shape and 6 more. The 10×50 s resolve 16 stars and the huge cluster is about half the field of view. The cluster is very large, very bright and not compressed. Using the big 11×80 binoculars will show 28 stars within the Coathanger. From a dark location, this cluster is easily seen naked-eye; the tail feathers in the arrow of Sagitta point directly at it.

NGC 7293

PN 22 29.6−20 48 6.3mag 15′×12′ The Helix nebula in Aquarius is the nearest and brightest planetary. Using the little 8×25 s I am quite amazed that the Helix is pretty easy in such small binoculars—a tiny but discernable dot. Moving up the big 11×80 binoculars makes it easy to see; it is the central 20 % of the field of view—a pretty bright, pretty large disk. With averted vision, the hole in the middle shows up about 5 % of the time and is never held steady. There are five stars, all pretty faint, which surround the nebula.

Chapter 16

How Can I Use a Computer to Help Me Enjoy Deep-Sky Observing?

Is this ever a loaded question! There are so many possibilities that are available to an astronomical computer user. The reason is that even a modest computer will provide much computational power and allow an astronomer to calculate, store and evaluate a huge amount of data.

This is going to be the chapter in the 2nd edition of this book that needs to be a complete re-write. Electronically, things in 2015 are very different than they were in 1999.

I am going to start out by telling you the tasks I do with a computer that have to do with deep sky observing. First, I have created a spreadsheet that has one row of data for each time I have been out observing. It has the date, location, telescope used, seeing, transparency, constellation observed and notes of who was there and information on what happened. The notes section can be anything from meteors seen to green flash to clouds moved in at midnight. As I write this I have just now entered notes on 1000 nights of observing. It is a trip down memory lane to look through that data to see the names of places I observed where I would like to return and old friends that I have not seen in many years. I can also see a reference to a telescope that I sold off decades ago or a rare night that I rated 10 out of 10. All of that is kept in one large Excel spreadsheet.

The notes which discuss what I actually saw at the eyepiece are in plain text files. I have created a set of files for each constellation and the typed in text of what I observed is in those text files in RA order. That order is pretty easy to maintain by NGC number because those numbers are in RA order. I just created a section of the file labeled "Non NGC" that contains the Abell, Barnard and other designations. So, once I know the constellation of an object I can look up my notes for that object by searching the text files. It made the original writing of this book simpler since I already had the notes in the computer.

S.R. Coe, *Deep Sky Observing*, The Patrick Moore Practical Astronomy Series,
DOI 10.1007/978-3-319-22530-2_16

This also makes searching the information much easier. I use an editor program called Ultraedit and the search function is powerful and easy to use. I just search for "bright" or "pretty faint" or "elongated" and the program finds those words within the text I typed in over the years. I know that some of you just don't want to take notes; you say it ruins the experience of viewing the sky. For me, the effect is the opposite. Having my notes available so I can remember when and were I viewed an object and what detail I saw really adds to my enjoyment of seeing the Universe in all its glory.

Planetarium Programs

There are a variety of programs that will bring the deep sky to your computer and provide a wealth of information about what is ready to be viewed tonight. Here are some freeware examples: HN Sky, Earth Centered Universe and Stellarium. Examples of programs which will cost you: The Sky, Guide and Sky Map Pro. There are more of each of these types and you may search for them on the Internet. I can recommend HN Sky and Sky Map Pro, I have many years of experience with them and they work well for me.

In general, these programs will provide you a map of the sky that looks much like a star chart. But that is only the beginning. A printed star chart generally provides the user the sky in one size or scale. So the author of the printed star chart chooses how large the sky and objects within it will be. This is the reason that the information I provided on printed charts gave several examples of different sizes of charts. Most of today's magazines about astronomy have a center section that contains a simple star chart. You can use it to find the brightest deep sky objects that are overhead this month. I also use that chart to answer questions like "which star in Aquarius is Gamma?".

All of the computerized star charts I have used can display a wealth of information about the objects they contain. Aim at the object you want information about and a right mouse click will show you more data than you probably need. If you are looking to find a planetary nebula, for instance, knowing its size is vital. This can give you a feel for how large or small the nebula will be. This can make finding it in a rich Milky Way field of view easier. A tiny nebula needs high power and exact knowledge of its position in that field of stars.

Knowing the magnitude of that galaxy you are searching for is also necessary. If you have a 6 inch telescope, or smaller, then you are somewhat limited in brightness. Even if you are the proud owner of a 10–12 inch scope then galaxies beyond 14th magnitude are going to be difficult. You are also going to have to save those faint objects for really good, dark skies. The good news about have a computer and one of these programs is that you can make you observing choices match the telescope and the evening of your observing session.

Among the galaxies I chose for the observations several of them have a low surface brightness. This means that they are faint in almost any telescope because the glow of the galaxy is spread out across a large area of the sky. Some nebulae also have low surface brightness, use low power and save them for a dark night.

Planetarium programs for your computer have the advantage of allowing you to zoom in or out depending on how you wish to display the sky. The next two figures will show what I mean and are one of the useful tasks I do with Sky Map Pro.

First, here is the entire constellation of Canis Major, which contains the prominent open cluster M 41.

Now, here is the same cluster. But I have zoomed in on it so I can see individual stars. Some of them have double star designations.

Fig. 16.1 Canis Major the entire constellation

Fig. 16.2 Zoomed in on cluster M 41

Creating an Observing List

There are a variety of ways to create observing lists. The easy one is to borrow from a source that has already done the work. The Astronomical League has a variety of lists that cover many types of observations. Many years ago I completed the Herschel 400 list and I am the holder of certificate number 71. Since William Herschel is a hero of mine I am proud of that achievement.

The Saguaro Astronomy Club has also posted a variety of observing lists and they are all free to download and enjoy. I am fortunate to have joined SAC many years ago and it is a group filled with deep sky observers and imagers. www.saguaroastro.org.

From the discussion above, you can use a planetarium program to create a list and a map of the positions of those objects. Just choose a constellation that is above the horizon tonight. If you don't know which constellations are up, then let the computer help you. Select the date and time you wish to observe and the display will be for that time. This assumes that you have set your location as well.

Now zoom in to the constellation you wish to observe. Virtually all these programs will allow you to choose the limiting magnitude of objects to display. To get started, pick something like 11th magnitude. Now the fainter objects will not be displayed. You now have an observing list and map of the positions for the objects you wish to observe. You can also click the mouse on an object and get more information on that object.

There are also programs that are made to the specific purpose of creating an observing list. I know of Sky Tools and AstroPlanner. I cannot speak directly about them because I have never used one of these programs very extensively. I can say that they allow you to type in information on your telescope, eyepieces and filters. Then you call up the objects you viewed and enter your notes. Once you have created that database the program will help you keep track of what you have observed and what you wish to view on a specific night.

Personally, I often use a spreadsheet as my database manager. The SAC database can be downloaded for free from the website given above. It is in a zip file in three formats. One is a simple text file, another is comma delimited so it can be easily entered in a database manager and the third format is as an Excel spreadsheet.

The good news about having it as a spreadsheet is that the information can be sorted by constellation, magnitude, object type or size. If I start with constellation sort I can pull out the information on all the objects in the SAC data that are in Scorpius, for instance.

Now, I can sort by magnitude and get rid of the faintest objects if I am using a modest aperture. Or I can find a few faint goodies if I have access to a friend's big scope. If there are open clusters I can go to the "number of stars" row and sort by that. Now I can get rid of clusters with less than 15 stars or so. Or I can get rid of planetary nebulae that are smaller then 5 arcsec. I hope that example will let you see that there are a variety of possibilities. I have no doubt that many of the programs mentioned above can do much of what I do in Excel. Pick one and get good at it.

Using the Internet

I have no doubt that there are many people who use the Internet for astronomy much better than I do. All I can tell you about are the places I visit. Just ask any astronomically inclined 15 year old teenager and they probably know of many other web sites to visit. You can spend a lot of time sitting in front of the computer.

I enjoy the site www.cloudynights.com, it is my astronomical home when I am on the Internet. It is a chat site for a wide variety of astronomical subjects. The moderators do an excellent job of stopping any “Meade versus Celestron” fights. There is plenty of great information available and many questions get answered here. If you look under “articles” and “monthly” there are 100 articles written by me that provide a set of observations of deep sky objects. In Europe, the Star Gazer’s Lounge is a similar site.

I would say that all of the manufacturers of telescopes, cameras and accessories have a web site at this point. They need one to keep even with the competition.

There are also plenty of Yahoo Groups that are filled up with users of a wide variety of equipment and software. Sign up is free and you can post your question or comment. Many times one of the persons who run the company will answer. Just go to www.yahoo.com and select “Groups”. Then enter “astronomy” for the search and watch what happens. There will be plenty of pick from the output of the search engine.

Folks, there is so much up on the Internet at this point that trying to write a printed book about what is there is a fool’s errand. So, just get good at a search engine or join one of the chat rooms and find what you need.

The other task that is easily done by spreadsheets is calculating. You can enter the formulae from the discussion on eyepieces into a spreadsheet. Then you can quickly and easily determine the magnification, field of view and exit pupil for a variety of telescope and eyepiece combinations. Modern spreadsheets also have the ability to make much more complex calculations, such as the exact position of a deep-sky object, given its RA and Dec and the time.

Mobile Computers (See Fig. 16.3)

Computing power is forever getting smaller and more portable. Astronomers, who must drive far from home to get to better skies, are making use of this technology to carry more electronic brain power into the field. The one thing you must do keep the screen at a low light level and red in color. This will let you maintain your night vision so that when to move to the eyepiece you will see faint stars and lots of detail.

I am going to be honest, now. I don’t own any of these devices. I am 66 years old and my phone flips open like a Star Trek communicator. Over the years I have gotten good at making a printed observing list and leaving myself room to write on that sheet of paper. Nothing I have seen makes me want to change. I am clear that I am becoming the minority.

Fig. 16.3 Carrying plenty of computing power into the field is getting easier

Astronomical Places to Visit on the Internet

www.saguaroastro.org—The SAC site, home of the SAC database. It is a free download and contains lots of data on the best and brightest 10000 objects to observe around the sky. Just go to the archive to get at the data. There are also many text files that contain data on Messier objects, double stars, the Best of the NGC and more.

www.ngcic.org—Home of the NGC Project. This site is the focal point of an effort to update and correct the data for all the objects which have an NGC or IC number. This is a daunting task, but one well worth the effort. If you have the time to go and observe some of the places where confusion still exists in this venerable catalog, then contact the NGC Project and join in.

www.astronomy.net—The Astronomy Net, a store-house of links to other sites. Plenty to explore, starting from this location. Virtually anything you can think about concerning amateur astronomy has a link from this site.

All of the most popular astronomy magazines have a large Internet web site. Just go to Sky and Telescope, Astronomy Magazine, Sky at Night or Astronomy Now and you will find lots of great information about what is in the sky for observing.

Chapter 17

Why Should I Set Up the Scope for a Public Viewing Session?

One of the most gratifying and fun evenings you can have with your telescope is a public viewing session. We have learned not to call this a star party, because the uninitiated public will show up expecting a "party". Providing people a chance to view the wonders of the sky for the first time allows you to get in touch with your first time, also. Do you remember the very first time you saw the rings of Saturn, or the Ring Nebula? Can you recall when a fuzzy globe focussed to become a giant ball of glittering stars? If that first evening has become a faded memory, then you can recover some of that sense of wonder by helping others get in touch with their place in the Universe (See Fig. 17.1).

On the other hand, if you are just starting to find your way around the sky this is a great opportunity for you also. Learn about two or three astronomical objects and be able to find them and then give the people nearby some information about what they are observing. By that, I mean about three or four sentences, not a speech! A crater on the Moon, whatever planet is up and one or two of the deep-sky objects at the end of this chapter will do fine. Believe me, compared with the average person, you *are* an expert.

First, make certain that your telescope is ready. It is exasperating for novices to try and use a scope that has too many peculiar features. Lubricate the focuser (if appropriate) so that it will deliver smooth motion. Align the finderscope; this will make it easier to locate objects, and that way you will appear like an accomplished observer. You want your audience to say, "Wow, how did you find that?"

S.R. Coe, *Deep Sky Observing*, The Patrick Moore Practical Astronomy Series,
DOI 10.1007/978-3-319-22530-2_17

Fig. 17.1 Take the opportunity to show the sky to a variety of people

Many times you will be setting up the telescope so that children can observe. This will demand some kind of ladder. The foremost thing to provide for them is safety. A rickety ladder is a very poor choice for a variety of reasons. Obviously, you don't want to see a child get hurt. Make certain that the ladder is tall enough, so that when an observer is at the eyepiece there is plenty of ladder above them to provide a grip for safety. A ladder that is so short that people have to balance themselves to get a look is a poor choice. Also, someone who is trying to keep from falling is unlikely to get a good look (See Fig. 17.2).

It is all too true that many people will point at the eyepiece and ask, "Do I look right here?" as they put the tip of their finger into the eyepiece! So expect too that at least a few of your eyepieces will become smudged. That is the reason I use old eyepieces for public viewing sessions. Just two eyepieces will generally be plenty, one that will provide between 75 and 100× magnification and another that gives 150–200×. Trying to show objects to novices at extremely high powers is often a waste of time. They don't see any more detail at 300- or 400-power and they find it much more difficult to get their eye in the correct position. Often an eye cup can be helpful, as it gives people a reference point.

The public will get the most from their first evening at the telescope if they have a knowledgeable guide. Spend some time learning about the objects you plan to show with the telescope, and know their distances, sizes and position in the sky. Often what is most interesting is the ability to provide some information concerning the physics of the object that is in the telescope. If you are planning on showing a planetary nebula, then be ready to speak about the fact that these are the end product of a star's life. If there are lots of galaxies overhead make certain that you are ready to explain the of vastness of what the telescope is presenting (See Fig. 17.4).

Fig 17.2 A ladder will be needed by some observers

Fig. 17.3 Comet Hale–Bopp is the main attraction at this public viewing

Fig. 17.4 If a bride can take time out to view through a telescope, anyone can. Be honest, if you had a photo of a woman in a bridal gown viewing with your telescope, wouldn't you include that in your book?

Even though this is a book about deep-sky observing, the planets will consistently be the objects that will stick in the mind of the first-time observer. So, make certain that you are ready to discuss the difference between the gas giants and terrestrial planets. Having a meteor streak overhead will also bring on many questions from the gathered crowd (See Fig. 17.5).

As you expound on the size, scale and complexity of Our Universe, be careful of the terminology you use. If you have been observing for some years, words that you use often are going to be quite foreign to lots of people. A few examples are: "light year", "globular", "nebula", "wide angle", "long focal length". Nothing is more confusing to a novice than the feeling that they need a decoding device to understand what is being said.

So, starting with autumn (for no good reason whatsoever) here is a listing of good objects to show at a public viewing session and a little info on what can be said about each of these constellations and objects within them:

Fig. 17.5 A variety of telescopes ready for a public viewing session

Aquarius

The Water Carrier. This place in the sky has always been associated with watery things. Ancient Babylonian art depicted a boy pouring water from an urn, while Arabians saw a two-handled water jar called an amphora.

M2

RA 21 h 33 m Dec –00 49 Mag 6.5 Size 13′ A globular cluster of at least 100,000 stars. About 50,000 light years distant. The cluster is 150 light years across. At the tremendous distance of this cluster, the Sun would be very dim at magnitude 20.7, only visible in the largest professional telescopes.

NGC 7009

RA 21 h 04 m Dec −11 22 Mag 8.3 Size 28″ × 22″ A planetary nebula. Called the "Saturn Nebula" by Lord Rosse because of extending arms or ansae which protrude from the nebula when seen under a good, dark sky. About 3900 light years distant, which means it is 0.5 light years across. This type of nebula will be formed by Our Sun at the end of its life, about 5 billion years from now.

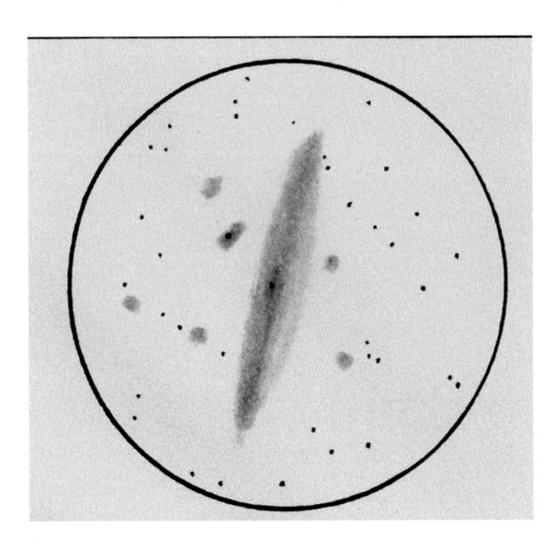

Fig. 17.6 NGC 7331 with a 13 in. telescope at 150×

Pegasus

Pegasus, the winged horse of Greek legend

M15

RA 21 h 30 m Dec +12 10 Mag 6.4 Size 12′ A globular cluster that contains at least half a million stars. It is 42,000 light years distant and about 130 light years across. Try some power; there are many beautiful star chains (See Fig. 17.6).

NGC 7331

RA 22 h 37 m Dec +34 25 Mag 10.4 Size 11′×4′ One of the brightest non-Messier galaxies. A large scope can show the dust lane. It can be seen in the finder or binocs. About 50 million light years distant. Several faint companions near.

Epsilon Peg

RA 22 h 43 m Dec +30 18 Mags 3/9 A lovely colored double star. A wide pair at 81″, they are easily split in any telescope. Tap the telescope tube and the dim star appears to swing around the brighter star. Fascinating!

Andromeda

the daughter of Cepheus and Cassiopeia. Andromeda is rescued by Perseus before she can be eaten by Cetus, the Sea Monster. All these people and animals are in the sky (See Fig. 17.7).

Fig. 17.7 The Andromeda galaxy is the closest spiral galaxy to our milky way

M31

RA 00 h 42 m Dec +41 16 Mag 3.5 Size 178′×40′ The largest, brightest *spiral* galaxy near the Milky Way. It is easily naked-eye from a dark site and in AD 950 was plotted on the stat charts of the Persian astronomer Al Sufi. 2.2 million light years distant. 150,000 light years across, about the size of the Milky Way. This is approximately what the Milky Way would look like from outside. Two companions nearby.

NGC 7662

RA 23 h 56 m Dec +42 33 Mag 8.6 Size 17″×14″ A very nice planetary nebula. I have always seen the color as blue or aqua. It is about 5600 light years distant, which means it is 0.8 light years across.

Almach

Gamma and—RA 02 h 04 m Dec +42 18 Mags 2/5 Means "The Foot" in Arabic, because it is the foot of Andromeda. A very nice double star; its members are 2 and 5 magnitude separated by 10″—I have always seen them as bluish and orange. About 95 light years away, so World War I had just been concluded as that light traveled toward Earth and your observers.

Fig. 17.8 The winter milky way from Cassiopeia to Perseus

Cassiopeia

The Queen. The mother of Andromeda and mother-in-law of Perseus. She was supposed to be very vain and was made to hang upside down over the North Pole because of it (See Fig. 17.8).

M52

RA 23 h 24 m Dec +61 35 Mag 6.9 Size 13′ An excellent open star cluster. It is 3000 light years distant and 10 to 15 light years across. A nice orange star is involved.

NGC 457

RA 01 h 19 m Dec +58 20 Mag 6.4 Size 13′ Another good star cluster. It includes Phi Cas, a 5th mag star. I have heard this called the Owl Cluster because the bright stars are like eyes of an owl with outstretched wings.

Eta Cas

RA 00 h 49 m Dec +57 54 A double star with color contrast. The stars are 4th and 7th mag and separated by 10″. I see the colors as light-yellow and orange. The two suns are 18 light years distant and about 68 astronomical units apart (an AU is the distance from the Sun to the Earth, about 93 million miles). They take about 480 years to complete one revolution about their center of gravity.

Fig. 17.9 The double cluster in Perseus

Perseus

The Hero. The rescuer of Andromeda and eventually her husband. He is pictured in the sky as holding the head of the Gorgon, the snake-haired woman, in his hand. This place is the location of Algol, the famous variable star (See Fig. 17.9).

***NGC 884** and NGC 869*

RA 02 h 22 m Dec +57 07 Mag 4.4 Size 60′ The Double Cluster is a unique and lovely pair of excellent star clusters that are so close together that they will fit in one wide field of view. The ancient Greek observer Hipparchus included it in a scroll he wrote in 150 BC. The clusters are about 8000 light years distant. That means that the ten brightest stars are about 60,000 times more luminous than our Sun. The Sun would be a magnitude 18 star and only visible in the very largest amateur telescopes.

M34

RA 02 h 42 m Dec +42 47 Mag 5.2 Size 35′ A nice open star cluster. It is 1500 light years distant and about 18 light years across. Easy to see in binoculars.

Fig. 17.10 Auriga includes the bright star Capella

Eta Per

RA 02 h 51 m Dec +55 52 A color contrast double star. The stars are 4th and 8th mag and separated by 28 arcsec. They are easily split at 100×. I have always seen them as gold and royal blue.

Auriga

The Charioteer. This honors Erechthonius, mythical king of Athens, who invented the four-horse chariot (See Fig. 17.10).

M37

RA 05 h 52 m Dec +32 33 Mag 5.6 Size 24′ One of the best winter open clusters. And telescope will show hundreds of members with several bright stars and beautiful dark lanes winding among the stars. About 4600 light years distant.

There is a lovely orange (or yellow) star near center that is not a cluster member—its motion differs from the cluster's.

M38

RA 05 h 29 m Dec+35 50 Mag 6.4 Size 21′ A nice open cluster. Has a cruciform shape at 100×. Look for NGC 1907, a cluster nearby in the Milky Way to the south.

Orion

The Hunter. He was fatally stung by Scorpius and put in the sky in a location opposite Scorpius, so that they are never above the horizon at the same time.

M42

RA 05 h 35 m Dec −05 23 Mag 4 Size 66′ × 60′ (See Fig. 17.11). The Great Orion Nebula was discovered only two years after Galileo invented the telescope. About 1900 light years away. Cleopatra was the queen of Egypt not long before the time when its light started to Earth. Density of the gas in this glowing nebula is a vacuum by laboratory standards, but enough material to make 10,000 Suns. The size is 30 light years across the nebula, which is 20,000 Solar Systems. Stars are being born within the nebulosity right now; our Sun hatched in such a cloud of material, over 4 billion years ago. There is a trapezium of four stars in the center, and over 50 variable stars are involved within the nebulosity.

Rho Ori

RA 05 h 13 m Dec + 02 55 Nice double star. Yellow and pale-orange pair are 5th and 9th mag, separated by 7″.

Fig. 17.11 There is much to show off in this shot of Orion at a public viewing

Iota Ori

RA 05 h 35 m Dec −05 57 One of the best triple stars in the sky. It is about 2000 light years away, all three stars are giants in size and luminosity. One companion is at 11″, the other 50″ away from the primary star. I have seen this triple as white, light green and purple. Honest.

Betelgeuse

RA 05 h 55 m Dec +07 24 Mag 0.7 variable This red star usually has its name translated "Arm of the Giant."

It varies its size over a period of about 5.7 years from 550 times the size of the Sun to 920 times Sol. This red super giant star is about 520 light years distant. It is one of the largest and most luminous stars visible to the naked eye. The luminosity varies from 14,000 times the Sun to 7600 times our Sun.

Taurus

The Bull that Zeus transformed himself into when he wished to carry off Europa, the eventual wife of the king of Crete. The constellation consists of two of the best open clusters in the sky, the Hyades and the Pleiades (See Fig. 17.12).

Fig. 17.12 The Pleiades or seven sisters are a nearby open cluster

Pleiades

—RA 03 h 47 m Dec +24 07 Mag 1.2 Size 100′ One of the best star clusters in the sky, M45 is named for the half-sisters of the Hyades. All had Atlas for a father. 410 light years distant. 10 light years across. 500 member stars. Three full moons fit across this cluster. Many lovely chains of stars. Our Sun would be a pretty insignificant star of 10th mag if at the distance of this group, so the brightest stars are all giants. The Japanese name of this star cluster is Subaru and there is a representation of the cluster on the back of every Subaru automobile.

Hyades

RA 04 h 27 m Dec + 16 00 Mag 0.5 Size 330′ The closest star cluster to Earth is about 130 light years distant. Aldebaran is *not* a member, but just happened to be in front in the same line of sight (See Fig. 17.13).

M1

RA 05 h 35 m Dec +22 01 Mag 8.4 Size 6′ × 4′ One of the few supernova remnants that can be viewed in a small telescope, the Crab Nebula is one of the most studied objects in the sky. Lord Rosse gave this object its name when he saw filaments within the nebula that reminded him of the claws of a crab. Chinese astronomers saw a bright star flare up in this location in AD 1054.

Fig. 17.13 The Hyades form the face of Taurus, the bull

That was the light from a supernova explosion, a large star ripping itself to pieces in an extremely violent explosion. The Crab is 6300 light years distant. There is a hot white dwarf star in the center that excites the gas to glow, just like a neon bulb.

Gemini

The Twins are Castor and Pollux, represented by the two brightest stars at the "head of the twins".

M35

RA 06 h 08 m Dec +24 20 Mag 5.1 Size 28′ A very nice open cluster. 2700 light years distant. 30 light years across. About 300 member stars. NGC 2158 is a compact cluster 30′ to the southwest in the Milky Way. Nice orange star near the center of the cluster.

NGC 2392

RA 07 h 29 m Dec +20 55 Mag 8.6 Size 47″ × 43″ One of the best planetary nebulae in the sky. It is large and bright for this type of object. 3000 light years distant and 0.6 light years across. Search for it at about 100× or so, you are looking for a gray-green dot. Then switch to high power (about 200×) to look for detail. Called the Clown Face or Eskimo Nebula because of dark features that can be glimpsed in the telescope at high power. There is a conspicuous central star on a night with even fair seeing.

Cancer

The Crab. Juno sent the crab to help the Hydra while it was in battle with Hercules (see below). The muscular brute stepped on Cancer and the sea-crab was transported to the heavens for trying his best.

M44

RA 08 h 40 m Dec +19 59 Mag 3.1 Size 95′ A large, scattered star cluster. It is named Praesepe or the Beehive. The cluster is 525 light years distant and about 13 light years across. So, the Black Death was getting a good grip on Europe about the time the light began its journey to your eyes. Several nice pairs and triples within the cluster at 100× or so (See Fig. 17.14).

Iota CNC

RA 08 h 47 m Dec +28 48 A nicely colored double star. The stars are 4th and 6th mag and separated by 31″. I have always seen them as gold and light blue.

Fig. 17.14 M-44 is the Beehive, a bright open cluster

Hydra

The Monster. The largest constellation in the sky has represented a variety of monsters. The most popular association is with the hundred-headed snake that lived in the swamp at Lerna until it was killed by Hercules.

M48

RA 08 h 13.8 m Dec –05 48 Mag 5.8 Size 42′ A large and bright open cluster. This Messier cluster was "lost" for many years until it was discovered that Charles Messier made a mistake in the declination of its position when he discovered it in 1771. The cluster is about 1600 light years distant and 20 light years across.

V Hydrae

RA 10 h 51.6 m Dec –21.3 This star varies in magnitude from 6.5 to 12 in a 533 day period. It is amazing in the telescope because it among is the reddest stars known. It is a carbon star, one of the rare class that shows strong bands of carbon molecules in its spectrum. An estimate of the distance to this star is about 1300 light years.

Leo

The Lion. This constellation represents the Nemaen Lion, which was killed by Hercules. He then wore the skin as a symbol of his prowess in combat. In ancient China, this constellation represented the Yellow Dragon.

Gamma Leonis

RA 10 h 19.9 m Dec +19.8 This is one of the most beautiful and best-observed double stars within reach of a telescope. It is a 2.1 and 3.5 magnitude pair that is separated by 4 arcsec. The name of this star in Arabic is Al Geiba, which means the Mane, its position in the body of the Lion. It is about 90 light years distant, which means the stars are 90 and 30 times the brightness of our Sun.

M66

RA 11 h 20.2 m Dec +13 00 Mag 8.9 Size 9′×4′ This galaxy is the brightest in the Leo subgroup M66 and M65 are both fine spiral galaxies and are about 38 million light years distant. Another galaxy, NGC 3628, is also in a wide field of view.

Ursa Major

The Large Bear. The most famous of northern constellations, this group represents Callisto, who was transformed into a bear by the jealous Hera, wife of Zeus. In Britain, it outlines Charles's Wain, the wagon used to transport King Charles I to heaven. The Big Dipper (known as the Plough in Britain and Ireland) is the outline most easily recognized here, and many of the stars in the Big Dipper have the same path through the Milky Way that Our Sun has. That means that the Big Dipper, Our Sun and a few other stars scattered around the sky form an open cluster that is gravitationally bound together (See Fig. 17.15).

Fig. 17.15 Most people know there are some dippers up there, somewhere. Point them out

Zeta UMa

RA 13 h 23.9 m Dec +54.9 This famous double star is named Mizar, which means Girdle or Loins, its position in the Big Bear. The first double star to be discovered, it consists of a 2nd and 4th mag pair separated by 15″. It is about 88 light years distant. There is a naked-eye companion, called Alcor. Mizar and Alcor form the "Horst and Rider"; they were used as a test of vision in ancient times.

M81

RA 9 h 55.6 m Dec +69 04 Mag 8.1 Size 26′ × 14′ This beautiful spiral galaxy is 38 acrmin from M82, a bizarre eruptive galaxy. They are about 7 million light years distant.

Canes Venatici

The Hunting Dogs. The names of the two hunting dogs are Asterion (Starry) and Chara (Dear).

Alpha Canes Venatici

RA 12 h 56.1 m Dec +38.3 Cor Caroli is this star's name, named for King Charles II of England; the name means "Heart of Charles". The components of this double system are 3rd and 5th mag and separated by 20 arcsec. At their distance of 120 light years the separation equals 770 AU. (AU means astronomical unit, the distance between Earth and Sun, about 93 million miles or 150 million kilometers.)

Therefore, the entire Solar System would fit five times between these stars. This has always been a lovely tinted pair in any telescope I have owned, the colors usually seen as blue–white and yellow.

M51

RA 13 h 30 m Dec +47 11 Mag 8.8 Size 9′ × 8′ The Whirlpool Galaxy is the standard example of a spiral galaxy. Its picture has graced the cover of many astronomical books over the year. It is 35 million light years distant, so no humans or human-like ancestors were alive when that light started toward your telescope. We live in a huge Universe.

M3

RA 13 h 42.2 m Dec +28 23 Mag 6.4 Size 6′ One of the very best globular clusters in the sky. It is about 40,000 light years distant and 220 light years across. Someone at Mount Palomar observatory *counted* 45,000 stars on a photographic plate taken there in the 1950s. The actual total number of members is about 1 million stars.

Fig. 17.16 The Coma star cluster is large, but an RFT or binoculars can show it off very nicely

Coma Berenices

Berenice's Hair in honor of Berenice II of Egypt. She cut her "golden tresses" as a sacrifice to Aphrodite when her king, Ptolemy III, returned safely from battle. The court astrologer told the royal couple that the golden locks had been transformed into a constellation and it has been included on star maps since the incident. Ptolemy III was king of Egypt from 246 to 221 BC (See Fig. 17.16).

Coma Star Cluster

RA 12 h 25 m Dec +26 Size 6° This large scattered star grouping is best studied in a pair of binoculars or a finderscope. The Coma star cluster is about 250 light years distant. So the British were just starting to overtax the settlers in the colonies when the light started from the cluster to your eye. The brightest members are 50 times as luminous as the Sun, so old Sol would be a 9.3 magnitude star at this distance, just visible in binoculars. There are about 80 members to the cluster.

NGC 4565

RA 12 h 36 m Dec +26 00 Mag 10 Size 15′×2′ This is the classic edge-on spiral galaxy. Its "flying saucer" shape and dark lane make it a lovely sight, it has also yielded many beautiful photographs. It is about 20 million light years distant and about 90,000 light years across.

24 Comae

RA 12 h 35.1 m Dec +18.4 A double star that I have always seen as blue and gold. It consists of a 5th and 6th mag pair separated by 20 arcsec.

Fig. 17.17 The Keystone is an asterism of four stars that form the centre of Hercules

Hercules

The Hero. This boisterous adventurer is the stuff of many Greek and Roman legends, including the voyage of the Argonauts and his Twelve Labors. He was placed in the heavens by Zeus at his death (See Fig. 17.17).

M13—RA

16 h 41.7 m Dec +36 28 Mag 6 Size 16′ One of the finest objects in the heavens, M13 is a large and bright globular cluster. This globular was discovered by Edmond Halley (yes, that Halley) in 1714. M13 is about 24,000 light years away and 160 light years across.

Scorpius

The Scorpion that stung and killed Orion. So, Jupiter put them into the sky 180° apart, that way Orion does not see the creature who slew him. Hawaiians see the Fish Hook of the god Maui at this point in the sky, placed there after he used it to fish the Hawaiian Islands up from the sea. The Chinese mark this celestial location as the Azure Dragon (See Fig. 17.18).

Fig. 17.18 Scorpius is a prominent constellation; the intruder is comet Halley

Alpha Sco

RA 16 h 30 m Dec −26.4 Antares means "Rival of Mars" because this red super giant star is near the average brightness of Mars and because it is the same ruddy color to the naked eye. Antares is about 10 times the mass of the Sun and at least 500 times the size 9000 times the brightness of Old Sol. A true super giant star by any standard. The outer layers of the star are very tenuous and would qualify as a laboratory vacuum. There is a 7th mag companion that is 3″ from Antares, making it a difficult split on nights of poor seeing.

M4

RA 16 h 23 m Dec −26 32 Mag 5 Size 25′ A very loose globular cluster that is easily resolved in most any telescope. Look for the curious "bar" feature of stars across the center of the cluster. It is about 6200 light years distant. So, the very earliest of Egyptian dynasties were being started along the Nile when the light started its journey.

M6

RA 17 h 40.1 m Dec −32 13 Mag 4.2 Size 15′ An open cluster that is bright enough to be naked eye under fairly dark skies. It is about 1500 light years distant and 20 light years across. There are about 80 cluster members. Look for the delicate chains of stars that form the "Butterfly" figure.

Fig. 17.19 The constellation of Scutum is a dense star cloud

Scutum

The Shield (See Fig. 17.19).

M11

Ra 18 h 48.2 m Dec −5 51 Mag 8 Size 9′ One of the richest open clusters in the Milky Way, M–11 consists of about 500 stars down to 14th mag. The Sun would be a dim 16th mag star at the 5500 light year distance of this cluster.

It is about 15 light years across. R.J. Trumpler calculated that an observer at the center would see several hundred 1st mag stars, the brightest 40 or so equalling or exceeding Venus!

Lyra

The Lyre. This stringed musical instrument was made from a turtle shell. When played by Orpheus it would cast a spell that charmed all the creatures of the earth (See Fig. 17.20).

Fig. 17.20 The constellation Lyra includes the bright star Vega

Epsilon Lyrae

RA 18 h 44.4 m Dec +39.7 This is the famous Double-Double. A pair of binocular or a finder will split the wide pair. Then each of those pairs will split in the main telescope at about 150×. The wide separation is 208″, then each tight pair is between 2 and 3 arcsec. All four stars are about 6th mag. The distance between the narrow pairs is about 165 AU, the size of the Solar System. The pairs are about 0.2 light years from each other.

M57

RA 18 h 53.6 m Dec+33 02 Mag 9 Size 80″×60″ The Ring Nebula is one of the most studied objects in the sky. It is certainly the best example of a planetary nebula. It is about 1500 light years away and half a light year across. The central star is very difficult to see in amateur telescopes. It is the nucleus of the star which ejected the material that formed the Ring itself. This dwarf star has a surface temperature of about 100,000°, much hotter than any normal star.

Vulpecula

The Fox. Originally Vulpecula et Anser, the Fox and Goose; maybe the Fox ate the Goose. In England it would be the name of a pub.

M27

RA 19 h 59.6 m Dec +22 43 Mag 7.3 Size 8′×5′ The Dumbbell Nebula gets its name from its shape. It is 900 light years distant and about 2.5 light years across. The central star probably released the gas which glows in the Dumbbell shape stating about 48,000 years ago. Lord Rosse used his 72 in. telescope to draw 18 stars involved within the nebulosity.

Collinder 399

RA 19 h 25 m Dec +20 11 Mag 4 Size 60′ The Coathanger is an open cluster that is large and bright. It is easily seen in a pair of binoculars or a finderscope. There is a curved line of stars that forms the hook of the Coathanger

Sagittarius

The Archer. Chiron placed an archer at this location in the sky to guide the Argonauts home after they had found the Golden Fleece (See Fig. 17.21).

Fig. 17.21 A wide-angle shot of the Milky Way with Sagittarius at the *bottom* and Aquila at the *top*

M8

RA 18 h 03.1 m Dec −24 23 Mag 5 Size 80′×40′ The Lagoon Nebula is a famous example of a diffuse nebula. There is a star cluster involved within the nebulosity. The name Lagoon comes from the dark lane that protrudes into the nebula. This object is about 4000 light years distant and 60 light years across.

M20

RA 18 h 02.3 m Dec −23 02 Mag 6.3 Size 28′ The Trifid Nebula is also named for the shape of the dark lanes that cut in front of the nebulosity. The Lagoon and Trifid may be sections of a vast nebulous cloud in that portion of Our Galaxy. So it is also at about 4000 light years; it is about 20 light years across.

M17

RA 18 h 20.8 m Dec −16 11 Mag 6 Size 45′×35′ The Omega Nebula, the Swan Nebula, the Checkmark—this object has been given several common names. It is about 5000 light years distant and 40 light years across. The bright "Checkmark" feature can be seen in any telescope, but use a UHC filter for the faint outer sections.

M22

RA 18 h 36.4 m Dec −29 54 Mag 5.1 Size 24′ This is an excellent globular cluster which is about 22,000 light years distant and at least 50 light years across. It is distinctly oval in shape.

M24

RA 18 h 17 m Dec −18 35 Mag 2 Size 120′×90′ The Small Sagittarius Star Cloud is an easily naked eye bright portion of the Milky Way. It is excellent in binoculars or an RFT. There are several dark nebulae that stand out on the north side.

M39

RA 21 h 32.2 m Dec +48 26 Mag 5 Size 32′ A bright, scattered open cluster that is best in binoculars or an RFT. There are about 30 members in the cluster. It is 800 light years away and 7 light years across.

Cygnus

The Swan. Zeus flew to visit the queen of Sparta as a swan and then placed the swan in the heavens to commemorate the event. This is also the Northern Cross (See Fig. 17.22).

Fig. 17.22 A wide-angle shot of the summer milky way

NGC 6826

RA 19 h 44 m Dec +50 31 Mag 8.8 Size 27″ × 24″ The Blinking Planetary is a striking sight. This planetary has a relatively bright central star and that makes for a unique show. As you look at the nebula directly the star overwhelms the nebulosity and it looks like a fairly bright star. Move your eye and look away from this planetary and the nebula is brightest, so the object grows in size. Looking back and forth will produce a "blinking" effect.

NGC 6960

RA 20 h 45 m Dec +30 43 Mag 7 Size 70′ × 6′ The Veil Nebula is a supernova remnant from a stellar explosion at least 30,000 years ago. It is about 1500 light years away and 70 light years across. This is the western part; it involves 52 Cygni. The UHC or O III filter works very well with object.

Beta Cygni

RA 19 h 30.7 m Dec +28.0 Albireo is one of the most famous double stars in the sky. It is easily split in most any telescope and has beautiful blue and gold color in most instruments. The 3rd and 5th mag stars are split by a wide 34″.

Albirea means "the Beak" because it is pictured as the beak of a south-flying swan (Cygnus).

Chapter 18

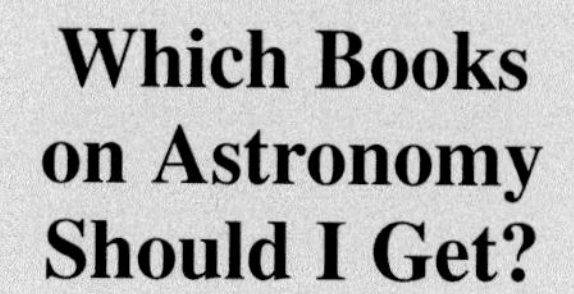

Which Books on Astronomy Should I Get?

If you decide to really become an accomplished amateur astronomer, then you will find that you start to collect books about astronomy.

A.J. has a method that appeals to me concerning how to decide if a book is worth purchasing. Look at the table of contents and read the introduction to see if the author is providing information that you might find worthwhile; Look through the index and see if you can find subjects easily. Find a subject that you know well and read those pages to see how it is explained. If the book passes all those tests, then it is worthy of consideration.

From my perspective, I found I had a lot of astronomy books just taking up shelf space. Several years ago, I went through them and asked: which of these do I really use on a regular basis? With that as the criterion, I put several dozen books in a box and sold them at the Saguaro Astronomy Club swap meet. I must admit that I used the money to buy an eyepiece that I had lusted after, but there was room on the bookshelf now.

In this chapter I am going to cover the books that I read often for data reference, observing ideas or just plain entertainment. Even though this became a pretty long list, as you will see, my purpose is to give you some information about what I have found useful and informative. This chapter will be organized by subject and I will supply the author, date, title, publisher and ISBN number for each book. For this second edition I will keep all the books from the first edition and add some I have found more recently. I know that some of the older books are out of print and you might need to search for them on the Internet or in a used book store.

S.R. Coe, *Deep Sky Observing*, The Patrick Moore Practical Astronomy Series,
DOI 10.1007/978-3-319-22530-2_18

Books for Novices (See Fig. 18.1)

Terence Dickenson and Alan Dyer (1991) The Backyard Astronomer's Guide. Camden House, Camden East, Ont. 0-921820-11-9

A great place to start. Lots of info on how the sky moves, finding your way around and how to use and appreciate a telescope and its accessories. Many of the questions that a novice astronomer will ask are answered in this book.

Kitchin, C.R. (1995) Telescopes and Techniques. Springer-Verlag, London. 3-540-19898-9

If you are just getting started in astronomy, this book is a good introduction at a modest price. Very useful information on sky coordinates, telescopes and mounts.

Moore, Patrick (1990) The Amateur Astronomer. W.W. Norton, New York. 0-393-02864-X.

Patrick Moore has been informing beginning amateur astronomers about how to enjoy the sky for half a century. His books are consistently informative, well written and very readable. This one contains much information about telescopes, observatories and what to see as the grand spectacle passes overhead.

Fig. 18.1 Books for Novices

Deep-Sky Observing (See Fig. 18.2)

Burnham, Robert (1978) Burnham's Celestial Handbook. Dover Publications, New York. 0-486-24063-0

My favorite deep-sky reference book. Includes lots of information on what to see for a large number of objects in the sky: variable stars, double stars, galaxies, nebulae, star clusters and famous stars are all included. The three volumes combine to form a massive work, arranged by constellation. The history and lore of each constellation and its principal stars is given in an easy to read manner. Some of the data (distances and magnitudes) are out of date, but that is not important. Robert Burnham demonstrates that Our Universe is a fun and fascinating place to be. My lifetime goal is to be able to say that I have observed every deep-sky object mentioned in Burnham's. I am over halfway there—just a few well-timed trips to Australia and I can do it!

Clark, Roger N. (1990) Visual Astronomy of the Deep Sky. Sky Publishing, Cambridge, MA. 0-933-34654-9

Lots of excellent information in this text. Roger Clark provides drawings of what he can see in a modest aperture. There is also much information about the human eye and what can be seen in a telescope by a persistent observer.

Fig. 18.2

There are magnitudes given for a variety of brightnesses in star clusters around the sky, so you can determine your limiting magnitude for the night. I refer to this book often.

Gilmore, Jess (2002) The Practical Astronomer's Deep Sky Companion Springer. 978-1852334741

This is a large book with great star maps that are guiding a deep sky observer to the best and brightest objects to view. There is information about each object and for imagers it gives an estimate of how long an exposure to give each deep sky goodie.

Murray Cragin, James Lucyk, Barry Rappaport (1993) The Deep Sky Field Guide. Willmann-Bell Richmond, VA. 0-943396-38-7

If you own the Uranometria star charts, you need this book. The authors have compiled data for the objects marked on each chart of the Uranometria. So, as you observe the galaxies, nebulae and clusters on the charts, this book can provide the magnitude, size and other data . An invaluable reference.

Hartung, E.J. (1968) Astronomical Objects for Southern Telescopes. Cambridge University Press, Cambridge. 0-521-31887-4

An excellent book about observing the sky and what can be seen in a variety of apertures. There is a northern addendum for the bulk of observers, but many objects in Sagittarius, Puppis and Canis Major are within reach of northern observers. The descriptions are well written and easy to use at the eyepiece.

George Robert Kepple and Glen W. Sanner (1998) The Night Sky Observer's Guide. Willmann-Bell Richmond, VA. 0-943396-58-1

Many descriptions, drawings and photos of deep-sky objects. Arranged by constellation in two volumes, this large work is the hardback version of the *Observer's Guide* magazine. The same two authors marketed Astrocards for many years and now this book is the compilation of those projects. Lots of great information on what to observe and how it will look when you find it. Your author was one of the contributors.

Chris Luginbuhl, Brian Skiff (1990) Observing Handbook and Catalogue of Deep-Sky Objects. Cambridge University Press, Cambridge. 0-521-25665-8

One of those books that you can't do without. The authors observed a wide variety of deep-sky objects and provide a very detailed observation, usually in several apertures. Many very useful detailed charts of galaxy clusters and other deep-sky objects.

Schaaf, Fred (1992) Seeing the Deep Sky. John Wiley, New York. 0-471-53068-9

A fine book that covers learning how the heavens are constructed. Many exercises, both indoors and out in the field, about the spectral classes of stars and the H-R diagram. Also some information and exercises on double stars and star clusters. This book does a fine job of teaching about the science of astronomy and how you can see it at work with your own eyes.

Star Charts

Tirion, Wil (1990) Bright Star Atlas 2000.0. WillmannBell Richmond, VA. 0-943396-27-1

An excellent simple star chart that plots stars to magnitude 6.5 and 600 deep-sky objects. This will do a fine job of showing a position for all the stars an observer could see on a clear night and all the bright and famous deep-sky objects. A great way to get started finding your way around the sky. With the *Bright Star Atlas,* a red flashlight and a pair of binoculars, an observer can keep busy for many nights.

Tirion, Wil (1982) Sky Atlas 2000.0. Sky Publishing Cambridge, MA. 0-933346-33-6

This mid-level set of charts show stars to magnitude 8.0 and 2500 deep-sky objects. This is a great set of charts and I used them for many years. I would highly recommend the Sky Atlas to deep-sky observers who are searching for a set of sky charts. There are three editions available: an unbound desk set of 26 flat charts, with black stars on a white background; an unbound field set, with white stars on a black background; and the deluxe version which is spiral bound and has a white background with color stars and deep-sky objects. Take the field edition out with you to observe and have the deluxe edition, just because they are so beautiful.

Wil Tirion, Barry Rappaport, George Lovi (1987) Uranometria 2000.0. Willmann-Bell Richmond, VA. 0-943396-14-X

These two volumes plot a third of a million stars and 10,000 deep-sky objects. It is obviously for astronomers who have found their way around the sky using the first two sets of charts. If you have a fair knowledge of what is available to observe and would like to step to a set of sky charts that you may never outgrow, then the *Uranometria* is for you. The two volumes overlap by a few degrees at the celestial equator, so consider them one giant book with two parts.

Computerized Astronomy

Ratledge, David (1998) Software and Data for Practical Astronomers. Springer-Verlag, London. 1-85233-055-4

You can spend many an hour searching the Internet for astronomy programs, or you can buy this book. It covers a lot of fine programs and includes a CD-ROM with lots of data, programs and images from the best sites on the Internet. There is also information on starting your own Web site and where to go for even more data and images. If you are like me and have an Internet connection that is slow, having some of these programs available on CD can be helpful (See Fig. 18.3).

Fig. 18.3

Telescopes and Accessories

Harrington, Phil (2007) Star Ware. John Wiley & Sons, New York. 978-0471750635

An overview of many telescopes, eyepieces, star charts and other astronomy accessories on the market today. The answer to "what should I buy?" is in here. If you are a gadget lover, this is the book for you. Also, an introduction to observing.

Mobberley, Martin (1998) Astronomical Equipment for Amateurs. Springer-Verlag, London. 1-85233-019-8

An excellent book that contains lots of information on subjects beyond the scope of this book. CCD cameras, using your computer to improve your astrophotos, video imaging and observatories. Lots of nifty gadgets, covered thoroughly and in a very readable manner.

The History of Astronomy

Ashbrook, Joseph (1984) The Astronomical Scrapbook. Sky Publishing Cambridge, MA. 0-933346-24-7

A compendium of interesting stories of past astronomers, observatories and astronomical happenings, taken from Joe's long-running Sky and Telescope articles. A great bedside book for amateur astronomers.

Couper, Heather and Henbest, Nigel (2009) The History of Astronomy. Firefly Books New York. 978-1-55407-325-2

I found this book very readable and contained good information on how we discovered so much about the Universe and how it operates.

Hathaway, Nancy (1994) The Friendly Guide to the Universe. Viking Press. New York. 0-670-839-44-2

A very well-written and just plain fun stroll through the story of how humanity has figured out the riddles of astronomy. Olbers' Paradox, early variable stars, Einstein's struggles with uncertainty and retrograde motion are explained from the point of view of the people trying to figure it out.

Krupp, E.C. (1983) Echoes of the Ancient Skies. Harper &Row, New York. 0-452-00679-1

A fascinating book on the methods our distant ancestors used to follow the sky and worship beneath it. A wide variety of cultures are discussed in an enthralling journey back through time.

Moore, Patrick (1992) Fireside Astronomy. John Wiley & Sons. New York. 0-471-93164-0

Another fine bedside book, it is composed of over one hundred short tales of astronomy. The vignettes cover everything from historical happenings and discoveries to modern observatories and exploration.

The Moon

Rukl, Antonin (1990) Atlas of the Moon. Kalmbach Publishing. Waukesha, WI. 0-913135-17-8

Yes, I know this is not a book about deep-sky observing. Hey, the Moon has to come up sometime! *The* standard lunar chart set. On the Internet, observers will refer to an observation of a lunar feature with "it's on Rukl # 23". Lots to see while waiting for the Moon to go away so you can look at deep-sky goodies.

Chapter 19

A Magical Evening

It certainly didn't look like it was going to be a great night under the stars, but A.J. and I were going out anyway! Clouds had built up all afternoon and it did not look good in the direction of the site for the Saguaro Astronomy Club Star Party, near Buckeye, Arizona. But, like I said, A.J. and I were going anyway. With A.J.'s trusty 8 incher in his van and my 6 in. RFT in the truck we got on the freeway and passed the time on the CB radio talking about computers, telescopes and plans for next weekend, when the Moon is New (See Fig. 19.1).

Arriving at the site, we find that two other club members have also decided to try their luck. As we munch on a sandwich, Jack Jones and Pierre Schwaar pull in. I set out the 6 in. as a sacrifice; if you don't set up a scope the clouds won't know that it is time to clear off. I find Albireo peeking through a small hole and we all comment it is the worst we have ever seen the colors of this famous double star. As it gets dark Rich Walker wheels his new truck next to mine and now we have seven, a number with magical powers. Like witches trying to cast a spell to make the clouds begone, we sit in a circle and discuss a wide variety of subjects: space programs, Galileo, Hubble Space Telescope, eyepieces, new films for astrophotography and observing sites for New Moon Weekend.

As Pierre swings his small RFT toward the west, a little clear sky starts to appear, we can see Jupiter. After Pierre mentions that he can see some detail on the planet, a few other folks start to get set up and I put an eyepiece in the 6 in. Jupiter is mushy, but getting better and as I look away from the eyepiece, the western horizon is starting to clear off. Time to get out the trusty binoculars to look for Comet Hale–Bopp. It is seen in the binocs, even with a light cloud cover and I point the 6 in. at it easily with the Telrad. The bizarre geometry that I have been seeing for weeks is evident right away, even at only 60×. The white core has a bright wedge

S.R. Coe, *Deep Sky Observing*, The Patrick Moore Practical Astronomy Series,
DOI 10.1007/978-3-319-22530-2_19

Fig. 19.1 My 13 inch Newtonain set up and ready

of material coming off it toward the north, but the tail is 90° to this bright region. A unique brightness contour for any comet I have ever seen.

As the evening progresses I make my way through a short observing list I have prepared. The Coathanger (Col 399) in Vulpecula is very nice. Using a 38 mm Erfle eyepiece that gives a 2° field in the 6 in., I can fit the entire Coathanger pattern in the scope and I count 38 stars in the cluster, which is not compressed, but has many bright stars.

The North America nebula is excellent as the clouds have really gone away now and the Summer Milky Way is very nice. With a 22 mm Panoptic eyepiece and a UHC filter, I can easily see the Mexico and Florida sections and the Pelican is somewhat faint, but it is there with direct vision.

Going over to visit with Rich Walker, we spend some time on Saturn at 220×, Cassini's Division is seen often in his 13 in. f/5 and there are three pretty faint satellites just off the tip of the rings. A quite good view, taking into account that an hour ago, we were just hoping for a hole in the clouds to steal a quick glimpse.

A.J. and I tear down about 1 o'clock and make our way back to our homes. I find I cannot sleep and after watching the late movie on Harry Houdini I step outside to see if the clouds have decided to stay away. It has become a sparkling clear night and at nearly 5 o'clock a lovely Crescent Moon is in conjunction with Venus and Mars. The Earthshine is obvious, cradled in the arms of the Moon and I decide to roll out my 10 in. f/6 for a closer look.

Venus is about 50 % illuminated and its beauty was best seen naked-eye. I make my way over to Mars and it is a tiny ball, but at 300× in the 10 in. f/6 I can just make out a small dark marking on the gibbous planet. Swinging over to the Moon shows me that several excellent features are near the terminator and lots of fine detail is seen using powers from 200× to 300×. Schroter's Valley is a snake winding

Fig. 19.2

its way out from crater Herodotus. Aristarchus is a crater with a very detailed floor and a prominent central peak. Both of these craters show places in the walls where material has slumped down in centuries past.

All in all, a truly magical night and one that got me back in contact with the reasons that I love to look at the sky. I enjoyed every kind of viewing on this night: low-power Milky Way scanning, the naked-eye beauty of the heavens, trying a variety of magnifications and filters on several deep-sky objects and a high-power look at the Moon and planets. Quite a memorable night. But it is night no longer. The Sun has risen, the day is heating up and I am going to try and catch up on my sleep. May your skies be clear. Goodnight all.

Appendix

Supplementary 110 Deep-Sky Objects

A variety of objects, some challenging, by the Deep Sky Group Saguaro Astronomy Club Version 1.0, dated Friday, 2 May 1991

This list is used by members of the Saguaro Astronomy Club as a supplement for deep-sky observing.

Object	Con	Type	RA (2000)	Dec	Mv	Size	Notes
NGC 404	And	E0	01 09.5	+35 43	10.1	4×4	Near Beta And.
Abell 262	And	GG	01 52.7	+36 09	13	120	5 galaxies in 13 inch.
PK47-4.1	Aql	PN	19 33.3	+10 37	13	161″×151″	Extremely faint in 13 inch.
B142-3	Aql	DN	19 40.7	+10 57	–	80×50	Best in RFT.
NGC 7492	Aqr	GC	23 08.4	−15 37	11.5	6.2	Tough in 13 inch scope.
IC 405	Aur	EN	05 16.2	+34 16	–	30×9	Flaming Star Nebula.
IC 342	Cam	SBc	03 46.8	+68 06	12	18×7	Low surface brightness.
King 14	Cas	OC	00 31.9	+63 10	9	7	20* in 13 inch.
IC 59	Cas	EN	00 56.7	+61 04	–	10×5	Paired with IC 63.
NGC 609	Cas	OC	01 37.2	+64 33	11	3	Use high-×to resolve.
Stock 2	Cas	OC	02 15.0	+59 16	4	60	Easy in binoculars.
IC 1795	Cas	EN	02 26.5	+62 04	–	27×3	Use nebula filter.
Mel 15	Cas	OC	02 32.6	+61 27	7	21	39* in 13 inch.

(continued)

S.R. Coe, *Deep Sky Observing*, The Patrick Moore Practical Astronomy Series,
DOI 10.1007/978-3-319-22530-2

(continued)

Object	Con	Type	RA (2000)	Dec	Mv	Size	Notes
Maffei 1	Cas	E3	02 36.3	+59 39	14	5×3	Much reddened, tough.
IC 289	Cas	PN	03 10.3	+61 20	12	45″×30″	15 mag central*.
NGC 7822	Cep	EN	00 03.6	+68 36	–	60×30	Large, faint, UHC.
Tr 37	Cep	OC	21 39.0	+57 30	5.1	50	30* involved in IC 1396.
IC 1396	Cep	EN	21 39.1	+57 30	3.5	50	Very large, use UHC.
IC 1470	Cep	RN	23 05.2	+60 15	–	15×1	Small comet shape
NGC 247	Cet	S-	00 47.0	−20 45	8.9	20×7	Low surface brightness.
IC 1613	Cet	Irr	01 04.8	+02 07	12	11×9	Member of Local Group.
IC 2165	CMa	PN	06 21.8	−12 59	12.5	9″×7″	Nearly stellar.
Sh 301	CMa	EN	07 09.8	−18 29	10.5	8×7	UHC filter helps.
Cr 140	CMa	OC	07 23.9	−32 12	4	42	Naked eye.
Mel 111	Com	OC	12 25.0	+26 00	2	275	Coma star cluster.
NGC 4676	Com	GG	12 46.1	+30 44	14.7	2.2×0.35	The Mice, use high-×.
Abell 1656	Com	*GG*	12 59.8	+27 59	11	120	Rich Galaxy Group.
NGC 5053	Com	GC	13 16.4	+17 42	9.8	10.5	Faint, very loose.
Abell 2065	CrB	GG	15 22.1	+27 39	14	30	Most difficult.
NGC 4038	CRV	Sp	12 01.9	−18 51	10.7	2.6×1.8	Use high-× for detail.
PK64+5.1	Cyg	PN	19 34.8	+30 31	9.6	5″	Campbell Hydrogen *
Ml-92	Cyg	RN	19 36.3	+29 33	11.7	8″×16″	Footprint Neb. Hi-×.
IC 1318	Cyg	EN	20 22.2	+40 15	–	60×30	Neby near Gamma Cyg.
PK80-6.1	Cyg	PN	21 02.3	+36 42	13.5	27″×18″	Egg Nebula.
IC 5146	Cyg	EN	21 53.4	+47 16	10	9	Cocoon Nebula.
NGC 6891	Del	PN	20 15.2	+12 42	10.5	15.5″×7″	Lovely aqua color.
NGC 6934	Del	GC	20 34.2	+07 24	8.9	2	40* resolved in 13 inch.
NGC 7006	Del	GC	21 01.5	+16 11	10.6	2.8	2* resolved in 13 inch.
NGC 4236	Dra	SBb	12 16.7	+69 29	10.7	23×8	Dim barred spiral.
Mrk 205	Dra	GL	12 21.7	+75 20	13	3.1×2.3	QSO, southwest edge NGC 4319.
NGC 1300	Eri	SBb	03 19.7	−19 24	11	6.5×4.5	Bar spiral in 13 inch.
Fornax Dwarf	For	Ep	02 39.7	−34 17	9	20	Low surface brightness.
NGC 1365	For	SBb	03 33.7	−36 08	10.5	10×5	For galaxy cluster.
IC 443	Gem	SNR	06 16.9	+22 47	–	50×40	Nr Eta Gem, use UHC.
J 900	Gem	PN	06 26.0	+17 47	12.4	12″×10″	PK194+2.1, use Hi-X.

(continued)

(continued)

Object	Con	Type	RA (2000)	Dec	Mv	Size	Notes
NGC 2371/2	Gem	PN	07 25.6	+29 29	13	74″×54″	Box shape at Hi-×.
PK205+14.1	Gem	PN	07 29.0	+13 15	14.1	10×6	Crescent shape, UHC.
NGC 7590	Gru	Sb	23 18.9	–42 14	12.5	2.7×1.1	Many galaxies near.
Abell 2151	Her	GC	16 05.1	+17 43	15	40	5 galaxies in 13 inch.
IC 4593	Her	PN	16 12.2	+12 04	11	12.5″×10″	Small, green in 13 inch.
NGC 2610	Hya	PN	08 33.4	−16 09	13	50″×47″	Small, comet-shaped.
PK238+34.1	Hya	PN	09 39.1	–02 48	13.4	268″	Low surface brightness.
NGC 3145	Hya	Sb	10 10.2	−12 26	12.4	3.3×1.7	Near Lambda Hya.
NGC 3309	Hya	GAL	10 36.6	–27 31	11.9	1.9×1.7	Abell 1060.
IC 764	Hya	S	12 10.2	–29 44	13.2	5×2	Low surface brightness.
NGC 5101	Hya	SBa	13 21.8	–27 27	12	5.5×4.9	Seeing the barred structure is difficult in 13 inch.
Leo 1	Leo	E3	10 08.4	+12 18	11.3	10.7×8.3	Low surface brightness.
Abell 1367	Leo	GG	11 44.5	+19 50	14	30	>30 galaxies in 1°.
IC 418	Lep	PN	05 27.5	−12 42	10.7	14″×11″	11 mag central*.
NGC 5897	Lib	GC	15 17.4	–21 01	8.6	12.6	Large, faint, loose.
NGC 2419	Lyn	GC	07 38.1	+38 53	10.4	4.1	No * resolved in 13 inch.
PK164+31.1	Lyn	PN	07 57.8	+53 25	14	400″	Faint, N2474 nearby.
NGC 2474/5	Lyn	EO	07 57.9	+52 52	13.5	0.9×0.7	Small, difficult to see in 13 inch.
NGC 2832	Lyn	E4	09 19.7	+33 46	13	3×2	Many galaxies near.
NGC 6703	Lyr	SO	18 47.3	+45 34	12.5	2.3	Two faint companions.
PK64+15.1	Lyr	PN	18 50.0	+33 15	13.3	17.5″×17″	Needs Hi-×, annular.
NGC 6791	Lyr	OC	19 20.7	+37 51	9.5	16	300*, large, faint.
NGC 2264	Mon	CL	06 41.1	+09 53	3.9	20	Cluster in Rosette.
NGC 2301	Mon	OC	06 51.8	+00 28	6	12	Blue/gold dbl* invl.
IC 2177	Mon	EN	07 05.1	−10 42	–	120×40	Eagle Nebula, difficult to see.
Mel 72	Mon	OC	07 38.4	−10 41	10	10	40*, many faint.
IC 4634	Oph	PN	17 01.6	–21 50	12	20″×10″	Small, light green.
B 72	Oph	DN	17 23.5	–23 38	–	30	The Snake, S-Nebula.
B 78	Oph	DN	17 33.0	–26 00	–	200	Bowl of the Pipe Nebula.
IC 4665	Oph	OC	17 46.3	+05 43	4.2	41	30* mags 7…

(continued)

(continued)

Object	Con	Type	RA (2000)	Dec	Mv	Size	Notes
J 320	Ori	PN	05 05.6	+10 42	12.5	11″×8″	13 mag central*.
Cr 70	Ori	OC	05 36.0	−01 00	0.4	150	Incl. Orion's Belt *s.
IC 434/B33	Ori	EN/	05 41.0	−02 24	–	60×10	B33 = Horsehead, small.
Sh2-276	Ori	EN	05 48.0	+01 00	–	600	Barnard's Loop.
Abell 12	Ori	PN	06 02.4	+09 39	13.9	37″	PK198-6.1, use OIII.
NGC 7317	Peg	GG	22 35.9	+33 57	15.3	0.7×0.5	Stephan's Quint, Hi-X.
NGC 7479	Peg	SBb	23 04.9	+12 19	11.7	4.4×3.4	Bar spiral, easy 13 inch.
Jones 1	Peg	PN	23 35.9	+30 28	12.7	314″	PK104-29.1, tough.
NGC 1275	Per	Pec	03 19.9	+41 30	12.5	3.5×2.5	Perseus A, small.
Mel 20	Per	OC	03 22.0	+49 00	1.2	185	Alpha Per Cluster.
NGC 1499	Per	EN	04 03.3	+36 25	–	145×40	California Nebula.
NGC 2818	Pyx	PN	09 16.0	−36 37	8.2	9	Edge of * cluster.
Scl Dwarf	Scl	E0	00 59.9	−33 42	10.5	75	Very low surface brightness.
IC 4628	Sco	EN	16 57.0	−40 20	–	90×60	N edge Tr24, UHC.
Tr 24	Sco	OC	16 57.0	−40 40	9	60	200* rich, =H 12.
IC 4637	Sco	PN	17 05.2	−40 53	13.5	21″×17″	Nearly stellar.
NGC 6337	Sco	PN	17 22.3	−38 29	12.3	38″×28″	High-×for detail.
H 16	Sco	OC	17 31.4	−36 51	10	15	Involved in Stinger.
NGC 6380	Sco	GC	17 35.4	−39 04	14	3.9	Faint, difficult to see in 13 inch.
NGC 6027	Ser	GG	15 59.2	+20 46	12.4	1.9×1.0	Seyfet's Sextet.
IC 4756	Ser	OC	18 39.0	+05 27	5.4	52	80* mags 8…
Sh2-84	Sge	EN	19 49.0	+18 24	–	15×3	Small Calif Neb, UHC.
IC 4997	Sge	PN	20 20.2	+16 45	11.3	2″×14″	Look for disk with UHC.
B 86	Sgr	DN	18 02.7	−27 50	–	4	Ink Spot, small.
NGC 6624	Sgr	GC	18 23.7	−30 22	8.3	5.9	Resolved in 13 inch.
NGC 6822	Sgr	Irr	19 44.9	−14 45	10	20×10	Barnard's Galaxy.
NGC 1432/35	Tau	RN	03 45.8	+24 10	–	30×30	Pleiades Nebulosity.
NGC 1554/5	Tau	RN	04 22.9	+19 32	var	30″	Hind's Variable Neb.
Mel 25	Tau	OC	04 27.0	+16 00	1	330	Hyades, binocs.
NGC 3172	UMi	GAL	11 50.3	+89 06	14.9	0.7×0.7	Polarissima, faint.
NGC 3132	Vel	PN	10 07.0	−40 26	8.2	84″×53″	Eightburst Nebula.
3C 273	Vir	QSR	12 29.1	+02 06	12	stellar	Brightest Quasar.
Cr 399	Vul	OC	19 25.4	+20 11	4	60	The Coathanger.
Stock 1	Vul	OC	19 35.8	+25 13	5.3	60	Easy in binoculars.
Sh 88	Vul	EN	19 46.0	+25 20	11.5	18×6	Elongated streamer.

Abbreviations: *B* Barnard, *Cr* Collinder, *H* Harvard, *IC* Index Catalog, *J* Jonckheere, *Mel* Melotte, *Sh* Sharpless, *Tr* Trumpler. Constellation abbreviations are the IAU standard. Type abbreviations: *GC* globular cluster, *OC* open cluster, *PN* planetary nebula, *EN* emission nebula, *SNR* supernova remnant, *RN* reflection nebula, *DN* dark nebula, *GG* galaxy group; galaxies are identified by their Hubble type. Sizes are in arc minutes, unless noted otherwise. *Mv* visual magnitude

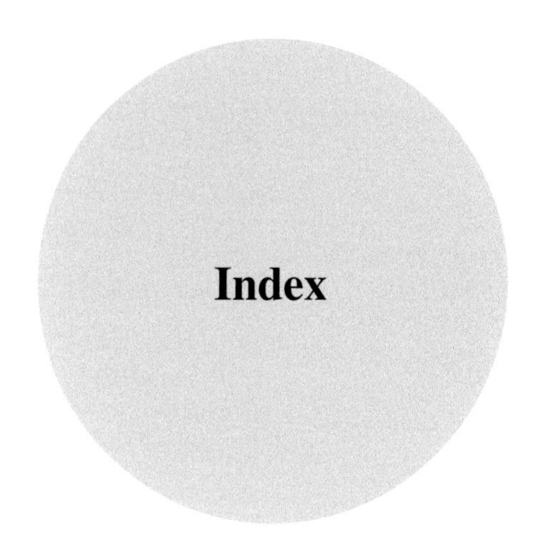

Index

S.R. Coe, *Deep Sky Observing*, The Patrick Moore Practical Astronomy Series,
DOI 10.1007/978-3-319-22530-2